THE ROLE OF TECHNOLOGY IN WATER RESOURCES PLANNING AND MANAGEMENT

SPONSORED BY
Task Committee on the Role of Technology in Water Resources
of the Water Resources Planning and Management Committee

Environmental and Water Resources Institute (EWRI)
of the American Society of Civil Engineers

EDITED BY
Elizabeth M. Perez, P.E.
Warren Viessman, Jr., P.E.

Published by the American Society of Civil Engineers

Library of Congress Cataloging-in-Publication Data

The role of technology in water resources planning and management / sponsored by Task Committee on the Role of Technology in Water Resources of the Water Resources Planning and Management Committee, Environmental and Water Resources Institute (EWRI) of the American Society of Civil Engineers ; edited by Elizabeth M. Perez, Warren Viessman, Jr.
    p. cm.
  Includes bibliographical references and index.
  ISBN 978-0-7844-1028-8
  1. Water-supply engineering--Case Studies. 2. Water quality management--Technological innovations. 3. Watershed management--Government policy. 4. Water resources development--United States--Planning. I. Perez, Elizabeth M. II. Viessman, Warren. III. Environmental and Water Resources Institute (U.S.). Task Committee on the Role of Technology in Water Resources.

  TD345.R65 2009
  628.1--dc22   2009012677

American Society of Civil Engineers
1801 Alexander Bell Drive
Reston, Virginia, 20191-4400

www.pubs.asce.org

Cover Photo: Banff, Canada. Courtesy of Elizabeth M. Perez, P.E.

Preface

This book is the culmination of three years of dedicated study by a Task Committee that was sponsored by the Environmental and Water Resources Institute within the American Society of Engineers. The goal of the Task Committee was to study the role of technology in water resources planning and management. The Task Committee included both experienced and emerging leaders in water resources planning and technology. Historical, current, and future trends in the use of technology to manage water resources are discussed through case studies. Summary sections are included in the first and final chapters of the book.

This book is intended to be a resource for students of water resources of all ages and abilities—from the executive water manager to the student at the university level. This book is a snapshot of the role of technology in water resources planning and management in 2008. This book is one of the most thorough discussions of this important topic and the authors hope that the book will inspire a new generation of water managers to use the best technology the field has to offer.

Acknowledgments

The information presented in *The Role of Technology in Water Resources Planning and Management* was obtained from numerous publications, web pages, and other sources. The authors are grateful for the support of the Environmental and Water Resources Institute and the American Society of Engineers for their support of this Task Committee. The authors are also indebted to those who authored the publications referenced in this publication and to those who were responsible for the material accessed on the Internet. The Task Committee would also like to offer our sincere thanks to Mr. Kyle Schilling, Dr. Eric Loucks, Dr. Timothy Feather, Dr. Wayne Losano, and our many supportive colleagues at Jones Edmunds & Associates in Gainesville—in addition to our many colleagues that reviewed individual case studies. References are documented at the end of each chapter and after each of the case studies presented to make them more easily accessible to those interested in a particular subject.

Task Committee Members

Stephen F. Bourne	PBS&J, Atlanta, GA
Kelly Brumbelow	Texas A&M University, College Station, TX
Hal Cardwell	United States Army Corps of Engineers, Alexandria VA
Neil S. Grigg	Colorado State University, Fort Collins, CO
Scott Knight	City of Aurora, CO
Daniel P. Loucks	Cornell, University, Ithaca, NY
Jayantha Obeysekera	South Florida Water Management District, West Palm Beach, FL
Elizabeth M. Perez	Brown & Caldwell, West Palm Beach, FL
Charles E. Shadie	United States Army Corps of Engineers, Mississippi Valley Division, Vicksburg, MS
Warren Viessman, Jr.	University of Florida, Gainesville, FL
Bill Werick	Culpeper, VA

Contents

CHAPTER 1

INTRODUCTION

Background

Whether proposed water resources policies and plans are implemented depends on their acceptability to decision makers. Final decisions are made politically, but they can be influenced by analyses made by water resources planners and researchers. The extent to which such studies are considered depends on: the credibility of the analyst, an understanding of the political and social climate of the planning region, and the directness of stakeholder input. Thus it is important to understand the role that technology has played and can be expected to play in supporting sound decisions regarding the allocation and protection of global water resources. Selected case studies covering a variety of settings and technologies provide an understanding of ways in which technology has supported water management policies and plans. Finally, a look to the future provides an insight into emerging technological approaches. See also (ASCE 1998; Dzurik 2002; and Loucks and van Beek, 2005).

Benefits of the Analysis

The authors believe that this publication will be a valuable reference for those engaged in water resources planning, management, and research. The book is not prescriptive. The intent is to show the role that technology has played historically, is playing at the outset of the 21st century, and is expected to play in the future. The authors hope that the information in this document will extend the knowledge of those charged with developing and managing the nation's water resources. The study does the following:

- Indicates the spectrum of technologies applicable to water resources planning, management, and policy making.
- Illustrates the emergence of new technologies such as adaptive management, shared vision modeling, and geographic information systems.
- Serves as a foundation for further exploration of similar topics.

Book Organization

Chapter 2 reviews the evolution of technology as applied to water resources planning and management. Applications of technology are illustrated in selected case studies in Chapters 3 through 5. Chapter 6 summarizes observations made during the study and presents expectations for the future. References appear at the end of each chapter, with the exception of Chapters 3, 4, and 5, where they are cited immediately after each case study. The modification to the standard reference format in Chapters 3 through 5 was made to conveniently refer the reader to additional information on each case study.

References

Task Committee on Sustainability Criteria (ASCE). (1998). Sustainability criteria for water resources," ASCE, Reston, VA., ISBN 0-7844-0331-7.

Dzurik, A. A. (2002). *Water Resources Planning*, 3rd edition, Rowman & Littlefield Publishers Inc., Savage, MD, ISBN 0-7425-1744-6.

Loucks, L. P. and van Beek, E. (2005). *Water resources systems planning and management: an introduction to methods, models, and applications*, UNESCO Publishing, ISBN 92-3-103998-9.

CHAPTER 2

EVOLUTION OF WATER RESOURCES TECHNOLOGY

Introduction

People have been planning, designing, and developing infrastructure and policies for the management and use of water for centuries, indeed arguably since the beginning of civilization. Few functions have been more important historically than managing water. Human demands and nature's supplies do not often coincide. Since we cannot live or function without water, engineering technology has been central to providing what we want and where and when we want it. Engineers have learned how to treat, control, and allocate water and then how to collect and treat wastewater before returning it to ground or surface waters. The multiple purposes of water— to meet agriculture, domestic, and industrial water supply demands; navigation; the production of hydropower; recreation—all depend on engineering technology.

During much of the 19th and 20th centuries, water resources planning and management activities were typically dominated by engineers and engineering technology. Dam building was an important component of traditional water-supply planning. Civil engineers had a major role in the construction and maintenance of the world's dams. An 1824 act of Congress established the United States Army Corps of Engineers (USACE) as the nation's preeminent water resources manager. Legislation passed in 1850 added water resources planning to the USACE's responsibilities, and in 1879 a Mississippi River Commission was established, with the USACE in charge of planning for an entire river basin. The USACE's interest in planning and managing the nation's waterways continues to this day in the form of numerous activities, including channelization and restoration projects, dredge-and-fill activities, harbor improvements, floodplain protection and management, and the construction and maintenance of a vast system of locks and dams on the nation's largest rivers.

In 1902, a second engineer-dominated federal agency—the U.S. Bureau of Reclamation—was created to deal with the physical and hydrologic conditions unique to the western United States. The Bureau's mission was to "reclaim" desert lands for agricultural and municipal uses. The Bureau, like the USACE, developed into a powerful planner and manager of water resources during the 20th century. A third federal agency, the Tennessee Valley Authority (TVA), was created in 1933 to integrate the use of all natural resources in the Tennessee River Basin.

Throughout most of the 20th century, the USACE, the Bureau of Reclamation, and the TVA used the concepts of conservation and multipurpose development to guide their planning of water resources projects. Conservation at the turn of the century meant using a scarce resource such as water to the fullest extent possible. It dovetailed with the multipurpose idea, in that the construction of a dam, for example, would not only provide flood control but would also store waters behind the dam for use as drinking water, for recreation, and for irrigation of crops.

3

The 1909 Rivers and Harbors Act authorized the agencies to also consider the provision of hydroelectric power in their planning. At the same time, policymakers stressed the need for comprehensive river basin planning as the best approach to the conservation of the nation's resources. In 1936 Congress mandated the USACE to employ a form of economic analysis known as *benefit-cost analysis* in its project planning. Other agencies followed suit.

In 1950, policymakers undertook a comprehensive review and analysis of water resources planning and management. Their report, known as *The Green Book* for the color of its cover, presented the classic economic efficiency model as the standard for analysis. Its revised version in 1958, *Proposed Practices for Economic Analysis of River Basin Projects,* covered the basic concepts of benefit-cost analysis, principles and procedures for project and program formulation, analysis of various project purposes, and cost-allocation methods. The report recommended that federal projects should not be undertaken unless their net benefits exceeded their costs. National economic development was the primary goal of federal project planning.

During the 1960s, environmentalists challenged many of the report's basic assumptions. Although classical economic analysis was not abandoned by federal water agencies, it was significantly modified in ensuing decades in recognition that economic development, urbanization, and population growth came at a heavy cost to the environment.

The traditional approach to water resources planning during this period was to identify needs for water and then devise a plan for meeting those needs. For water supply projects, the typical approach was to forecast future water demands and then try to find and deliver the water necessary to meet those demands. For purposes such as navigation, hydropower, flood control, or any other purpose or combination of purposes, the same approach was taken, as long as the value of economic benefits exceeded its costs. Often little if any attempt was made to integrate supply management and demand management options. Traditional planning tended to be relatively narrowly focused and exclusionary. Once an agency had committed itself to a plan, little room was left for altering the chosen course.

Comprehensive river basin planning studies carried out largely by the USACE or the Bureau of Reclamation were often excellent exercises in judgment but not based on very thorough analyses of multiple options and impacts. These studies were very visible and occasionally resulted in the construction of large dams and changes in policy.

Municipal water supply utility planning rarely involved the public-at-large, outside experts, or government regulators. Demand estimation and the assessment of supply alternatives took place within the utility (or a single planning unit within the utility); only the final product was made available for review or regulatory approval. Major investment decisions were often made with little or no oversight or concern for the larger regional impacts if any.

One can argue that part of the reason for these traditional approaches to planning and management is that they were the best that could be done with the technology that existed. It was not easy to integrate multiple disciplinary inputs into an analysis and generate and evaluate multiple alternatives in a search for the best compromise. Neither the needed computer hardware nor software was available. When this technology began to appear in the early 1960s, planners and managers had neither the training nor confidence to use them. It has taken time to get to where we are today, and yet there is still a noticeable gap between the state of the technology and the state of its use.

Water Resources Technology: Where We Have Been

Technology for water resources planning, design, and management has been dominated by modeling. Models provide a means of predicting what the impacts of a particular design or operating policy might be before investing substantial funds to implement the design or policy. Models can also explore alternative designs and policies in a search for those considered most acceptable.

Before the mid 1950s, most modeling efforts focused on the design and use of physical models, often constructed in hydraulic laboratories. Since the advent of the analog and then the digital computer there has been a gradual but marked change in the education and practice of water resources planners and managers. The discipline is now dominated by the development and use of predictive models designed to be solved on digital computers.

Early computer models of the 1950s through much of the 1980s were designed for main-frame computers. IBM punch cards and occasionally paper and magnetic tapes were used to input programs and data—and paper was used for tabular outputs of model results. Those main-frame computers had only a tiny fraction of the speed and capacity of today's laptops, but they provided a first opportunity to broaden the scope of planning and management studies. Many comprehensive master plans for river basins were developed with the aid of such computer technology. Each computer simulation could take days to run, only to find that some mistake had been made in the program or data. For the younger readers of this book, that may be hard to imagine. The tabular information then had to be converted into useful graphical displays by hand.

Eventually, computer display terminals replaced punched cards. Then came the microcomputers – Apple and then IBM, Digital, and HP to name a few. Perhaps of equal importance were the graphical display terminals. Modelers could display outputs graphically. Then we learned how to use a digitizing pen and tablet and later a mouse to input data directly to the display terminal. We eventually got color display capabilities and could try to communicate data in five dimensions: x, y, z space, time, and color. Such displays were useful, for example, to show changes in water quality (represented by different colors) in three-dimensional water bodies (reservoirs, lakes and aquifers) over time. Some of these displays were the predecessors to the geographic information systems (GIS) many of us use on a daily basis.

Then came menu-driven models, GIS, and the development of computer-based decision support systems (DSS) of various types. This shift is characterized by viewing water projects in an integrated manner as opposed to considering them discrete elements. Textbooks usually define GIS as a combination of hardware and software that allows data to be managed, developed, analyzed, and maintained in a spatial context. GIS has also been defined in many other ways that are perhaps more helpful in understanding what it is and how it can successfully be applied to water resources projects. At a 1998 conference, Jack Dangermond, the President of Environmental Systems Research Institute (ESRI), noted that GIS is a visual language, a framework for studying complex systems, integrating our knowledge about places, and helping us to organize our institutions (Shamsi 2002). We can also add to this definition that GIS helps us do a better job of managing water resources, enhances the life of the public, increases efficiency, decreases time spent on repetitive tasks, and further ensures the success of ecological restoration efforts.

During this development and increasing use of digital computers, and especially the ability to display spatial time-varying data pictorially, physical modeling has been on the decline. While physical models are still considered essential for complex design studies, they have become less important for planning and management. This book will focus primarily on the impact computers and their associated technologies had and will have on our planning and management activities.

Accompanying this rapid increase in the capacity of computer-based technology has been the development of mathematical and computational tools that permit a systems approach to water resources planning and management. The systems approach focuses not just on the design and operation of individual components of a multiple component system—whether it be a single water treatment plant or an entire river basin consisting of multiple water and wastewater treatment plants, reservoirs, hydropower plants, diversions, recreational facilities, and the like— but of the system as a whole. The focus is on the maximization of the performance of the entire system, not just of each component. This systems technology has also allowed the explicit inclusion of multiple disciplines within such analyses. We are no longer basing our recommendations regarding building or not building dams or levees on the availability of good construction sites and on economic efficiency criteria alone, but on numerous inter-disciplinary performance criteria. Today models often identify efficient tradeoffs among a host of economic, ecologic, and social performance criteria. As our technology and knowledge improve and as society's goals change, so do our planning and management models for identifying and evaluating alternative water resource systems that meet those goals.

The Harvard Water Program and its book *Design of Water-Resource Systems* (Maass et al. 1962), can take a large share of the credit for changing the discipline of water resources planning and management. The authors of that book were showing the rest of us how one could simulate complex multi-component systems using digital computers and how to evaluate alternative designs and operating policies based on economic criteria. They demonstrated how optimization and statistical methods could help us identify good alternatives to simulate. Furthermore, we could even quantify some of the uncertainties associated with our analyses.

To the credit of the authors of Harvard's design book, much of what they wrote still applies today even though the technical and political environment in which planning and management takes place is quite different than it was in the late 1950s and early 1960s. There are more than just economic criteria to deal with today and the technology they were excited about has surely exceeded what any of them would have guessed would take place in the following half century. They recognized that water resources planning and management take place in a political environment, yet they did not have what we have today to facilitate the interaction between modelers and those making decisions, nor the transfer of model results to the stakeholder-driven political process.

Water Resources Technology: Where We Are

Today the Bureau of Reclamation no longer considers itself a construction agency, but instead a management and planning organization that employs watershed management and river basin planning to help states and the private sector meet *all* water needs of the arid but highly populated West. The USACE continues to be a construction and engineering agency, but is also pursuing a number of more environmentally sensitive programs such as wetland protection and restoration, mitigation banking, floodplain management, and watershed planning. Both agencies hire biologists, economists, geologists, anthropologists, lawyers, and individuals with other applicable disciplinary backgrounds. And both agencies, like most water management agencies at the federal and state levels and even at local levels, employ multi disciplinary modeling to obtain the information they need to make informed development and management decisions.

Today's models are an essential part of any planning process, whether focused on flooding problems, reservoir operation, groundwater development, water quality, ecological restoration, or water allocation. Models and their computer technology can address a range of complex water resources and environmental problems from hilltop to ocean in an integrated fashion. Models of varying complexity, and thus of varying data requirements, are available. Their choice will depend in part on the needs of any planning or management study as well as the data and time available for the study.

Numerous models are available today to predict the runoff from watersheds due to precipitation. More complex models can also include the sediment, nutrient, and other pollutant loads in that runoff. This runoff and its constituents, as appropriate, can enter surface water bodies and/or groundwater systems. Models can predict the interaction between ground and surface water bodies, and the flows and their constituents in stream and river channels. These routing models can be based on simple mass balance and advection-dispersion relationships, or they can be based on hydrodynamic computations. They can also include ecological components of aquatic systems.

Models are available for the study of reservoir operation, flood forecasting and control, storm-surge impact prediction, embankment erosion, dam break planning, and for ecosystem restoration. They can also be used for real-time operation and management.

The complexity of many river basin systems that include multiple reservoirs and demand sites would benefit from the use of models for periodically informing managers about how best to manage such complex systems based on selected criteria. While this is currently done in some basins, it often is rejected in others because of institutional or political considerations. Where water is scarce and where there can be conflicts, water management policies for future operations are often resolved in the courts. Rarely do such decisions take into account how each water user in a complex multi-reservoir system can impact all other users downstream and thus how their allocations should be based on the existing storage volumes in downstream reservoirs. Considerable efficiency is lost through such a process. Until modelers can inform attorneys of how such real-time management models could work in specific cases in ways they can accept, many opportunities for efficient water management during stressful periods will be lost. Nevertheless, models exist for such applications. The argument for real-time water management applies during periods of flooding where multiple reservoirs, release basins, and levees can be used to reduce the potential damage.

Models also exist that predict changes in the river bed and planform, including bank erosion, scouring, shoaling associated with, for instance, construction works and changes in the hydraulic regime. This involves the transport of multiple sediment sizes ranging from fine cohesive material to gravel. Such models have been applied for morphological studies in small- and large-scale rivers, meandering as well as braided, from steep mountain rivers to estuarine environments as well as for reservoir sedimentation, simulating time scales from just a few hours to several decades.

If in-stream water quality processes are of concern, there are also a variety of water quality simulation packages that can be used. These types of models predict the transport and interaction of pollutants such as nutrients and oxygen depleting constituents, and their effects on algae and thus dissolved oxygen, the parameter often used to measure stream health. Similarly, toxins and heavy metals may also be simulated. Models may use complex kinetic reactions to simulate constituent interactions in the water, soil, and air, or may apply simpler empirically-derived reduction coefficients to constituents. Water quality modeling is an important step in setting up total maximum daily loads (TMDLs).

For less-detailed but more-comprehensive river basin infrastructure design and management, planning models are available and often are packaged within an interactive menu driven interface. This interface typically allows the display of maps or pictures of the area of interest, over which a node-link representation of the water resource system is defined. All site-specific assumptions are included in the input data. Such models provide a simple framework for managers and stakeholders to address multi-sectorial allocation, discharge, and water quality issues in a river basin. They can represent all elements of river basin systems, including rainfall-runoff, surface water, groundwater, reservoirs, hydropower, various users, water and wastewater treatment plants, and ecosystem parameters, as appropriate in specific studies. They are often used to facilitate communication with non-technical audiences.

Once such comprehensive yet preliminary screening models identify what should be simulated in more detail, more complex models can be applied. For a more detailed look at hydrologic processes in watersheds, for example, integrated hydrologic models can be applied to the entire land phase of the hydrologic cycle. This could include the use of a three-dimensional, numerical groundwater model together with numerical models for overland flow, unsaturated flow, solute transport, agricultural practice, and evapotranspiration. All this could be coupled to urban watershed models that include gutter and storm sewers, if applicable.

Geographical information systems (GIS) are becoming increasingly a part of many planning and management models. Increasingly, hydrologic models are directly linked to GIS data bases. Such tools are useful for two- or three-dimensional spatial calculations such as delineating watershed boundaries and stream and river paths, defining drainage areas and the areal extent of any other data layer, and modeling distributed runoff. In the future, hydrologists will increasingly rely on GIS data and standardized ways of describing those data so that they can be used consistently and efficiently to solve a wide variety of water resource problems at any spatial scale.

Decision Support Systems and Shared Vision Modeling

Planning and management activities today are often participatory processes involving input from many stakeholders. Typically these stakeholders have multiple interests and multiple goals and needs. Working in this multi-stakeholder multi-objective arena is not as easy as engineering design, even though the latter may require much more specialized training. Our planning and management models have adapted to this environment.

Modern water resources planning and management involves negotiation and compromise. So how do we model to meet the information needs of all stakeholders? How can we get them to believe in and accept these models and their results? How do we help them reach a common or shared vision? How do we get the information derived from our modeling technology entering the political debate about when and where to do what and why?

9

To be useful in the political decision-making process, the information we generate with all our models and computer technology must be understandable, credible, and timely. It must be just what is needed when it is needed. It must be not too little and not too much.

How do we know what is the right amount of information, especially if we are to have that information available and in the proper form, before, not after, it is needed? Obviously we can't know this. However, over the last two decades or so this issue has been addressed by developing and implementing what is called *decision support systems* (DSSs). These interactive modeling and display technologies can, within limits, adapt to the level of information needed and can give decision makers some control over data input, model operation, and data output. But will each decision maker and each stakeholder trust the model output? How can they develop any confidence in the models contained in a DSS? How can they modify those models within a DSS to address issues the DSS developer may not have considered? One answer to this has been to involve the decision-makers themselves not only in interactive model use but in interactive model building as well.

Involving stakeholders in model building accomplishes a number of things. It gives them a feeling of ownership. They will have a much better understanding of just what their model can do and what it cannot do. If they are involved in model building, they will know the assumptions built into their model. Being involved in a joint modeling exercise is a way to better understand the impacts of various assumptions. While there may not be agreement on the best assumptions to make, stakeholders can learn which of those assumptions matter and which do not. In addition, just having numerous stakeholders involved in model development will create discussions that will lead to a better understanding of everyone's interests and concerns. Through such a model building exercise, those involved may not only reach a better understanding of everyone's concerns, but also a common or 'shared' vision of at least how their environmental system (as represented by their model) works. Experience in stakeholder involvement in model building suggests that such model building exercises can also help multiple stakeholders reach a consensus on how their real system should be developed and managed.

Operation Management Modeling

It is not uncommon for operators of regional water resource systems such as those in the Columbia, Missouri, or Tennessee Valley river basins to be managing multiple reservoirs, with competing demands for the water including but not limited to water supplies for irrigation, domestic consumption, hydroelectric generation, recreation, transportation, and the preservation of habitats and species. In this situation, real-time decision support systems can again be helpful.

Operations management involves continual communication between project stakeholders and those responsible for water management. Communications

10

technology and better models for facilitating citizen participation promise to increase the responsiveness of water management agencies to changing stakeholder objectives and goals.

Hydroinformatics

Hydroinformatics is a term, originating in Europe, given to the link between computer models of water resource systems and displays that enable the effective communication of model results to those who need that information. This area of water resources technology grew from the need to display the results of computational hydraulics models in more understandable ways. The numerical simulation of water flows and related processes remains a mainstay of hydroinformatics, but this has broadened to an interest in the use of artificial intelligence techniques such as artificial neural networks, support vector machines, and genetic algorithms and genetic programming. These methods might be applied to large collections of observed data for data mining for knowledge discovery or with data generated from physically based models in order to generate computationally efficient emulators of those physical (and often computationally demanding) models for some purpose.

Hydroinformatics recognizes the inherently social or political nature of the decision-making processes in water management and strives to understand and meet those needs. In this sense this term represents the same activity and goal as those who are active in the development of interactive, graphics-based decision-support systems. Both hydroinformatics and those involved in development of water resources technologies strive to support comprehensive water resources planning and management decision making at all levels of governance and at all levels from broad regional planning to local operations management.

Planning and Management Modeling

Current modeling technology and education of professionals in the use of this technology permits individuals in such agencies as the USACE, the Bureau of Reclamation, and the TVA—as well as numerous state and local water resources planning agencies to undertake studies in the following:

- Water supply availability and reliability applied to surface and ground waters taking into account basin development projections, administrative and legal requirements, and constraints related to conveyance and reservoir operations, aquifer pumping, and water demand redistribution.

- Water use assessments of past and present water use for determining rates of water consumption and water loss in water supply distribution systems and on-farm water application practices, a prerequisite for identifying opportunities for improved water use efficiency and validating water requirements and historic beneficial water usage in water rights proceedings.

11

- Water diversion requirements of present and projected consumptive and non-consumptive water needs including domestic, commercial, industrial, rural, irrigation, municipal, energy, and environmental water use sectors for determining storage and distribution facility sizing requirements that meet the needs of the water service area.

- Watershed hydrology_investigations of precipitation–runoff relationships, determinations of runoff in ungauged watersheds, studies of surface water–ground water interactions including baseflow separation, determinations of reconstructed streamflows for past and undepleted streamflow conditions, and analyses of aquifer conditions including recharge characteristics, transmissivity, and safe yield.

- Water resource investigations for improved river and reservoir administration and operation, conjunctive use of surface water and ground water supplies, new and expanded storage and conveyance facilities, water banking, and water exchanges and transfers are a few of the water supply augmentation options that can be examined.

- Design of advanced water resource analysis tools, including simulation models and relational database management systems, and evaluation of the potential of new and improved concepts, methodology, and technology and work on the advancement of new analyses techniques for improving water resources management.

- Development of improved drainage and soil quality and design alternate sprinkler, surface, and precision irrigation practices and strategies to improve crop productivity and quality, irrigation use efficiency, and application efficiency.

- Development of improved crop production systems that optimize yield and quality, improve water, fertilizer and pesticide management, and reduce within-field variability by incorporating precision technologies as well as the advances in irrigation scheduling and application.

References

Maass, A., Hufschmidt, M.A., Dorfman, R., Thomas, H.A., and Marglin, S.A., Fair, G.M. (1962) *Design of Water Resource Systems: New Techniques for Relating Economic Objectives, Engineering Analysis, and Governmental Planning*, Harvard University Press, Cambridge, MA.

Shamsi, U.M. (2002). *GIS Tools for Water, Wastewater, and Stormwater Systems*, ASCE Press, Reston, VA.

CHAPTER 3

WATER SUPPLY CASE STUDIES

This chapter addresses one of our most basic concerns as human beings—water supply. As previously mentioned, the need for water has historically driven humans to develop better technologies and is perhaps one of the most dominant forces in shaping society. It is with this basic need in mind that we grouped these case studies on Washington, D.C., Texas, and Libya. While all of the case studies were driven by basic water supply needs, the outcomes and the involvement of technology are very different. The first case study, on Washington D.C., highlights an important historical case in which regional water supply planning was used. The Washington D.C. case was one of the first regional solutions to water supply concerns and it demonstrates a situation where technology was important but not central to the solution. In the cases of Texas and Libya, technology played and is playing a more central role, although the cases are very different. The Texas and Libya case studies are excellent examples of the use of modern technology to address important and urgent water supply needs. These case studies also note important themes related to the importance of stakeholder and public involvement, as well as the ability of drought to drive the development of technologies in times of need.

Washington D. C. Metropolitan Area Water Supply

Introduction

Since the early 1600s, when humans first settled in the Potomac River Basin, there has been a continuing concern with the river and its use. For about the first two hundred years, the focus was mainly on transportation and communication. Following an era of western movement, attention shifted to issues related to water supply and later to pollution control. The driver of these changes was the ever-increasing population of the basin, particularly in the Washington, D.C. Metropolitan Area (WMA).

Shortly after the turn of the 20th century, water resources planning on various scales became commonplace. Hundreds of studies were generated, some by the United States Army Corps of Engineers (USACE), others by state and local levels of government. But most of these products became residents of library shelves rather than blueprints for action. A major reason for inactivity was the inability of the many state, federal, local government, water utility, and other stakeholders to agree on common solutions. In 1981, Dr. Abel Wolman, aptly put it: "An orderly management of the array of functions [uses] … is distinguished by its absence. That it is needed has generally been agreed upon…Suspicion of an overlying authority by whatever name, has characterized public and private reactions" (Wolman 1981).

Institutional conflict characterized most early attempts to implement proposals for water supply management in the Potomac River Basin and resulted in a stagnation of effort. But concurrent with this inactivity, pressure was building to "do something" as the metropolitan area population grew and occasional low flows in the river became increasingly troublesome.

The WMA water supply problem centered around the fact that three water supply agencies—the Washington Suburban Sanitary Commission (WSSC) for Prince George's and Montgomery Counties in Maryland; the Fairfax County Water Authority (FCWA) for Fairfax county in Virginia; and the USACE Washington Aqueduct Division (WAD) for the District of Columbia— all depend on the Potomac River to satisfy the water demands of their service areas (McGarry 1990). The seriousness of the problem was divulged by studies which showed that during previously recorded droughts, average daily flows in the Potomac River were significantly below contemporary and projected peak daily demands. During the drought of 1966, for example, a one-day low flow of 1.47 million m^3/day (388 mgd) was the worst on record (Sheer 1981; Hagen et al. 2005). That was the first time the river flow had ever fallen below the maximum water supply withdrawal rate.

With a rapidly increasing population, but relatively stable river hydrology, droughts of a magnitude equal to or greater than those such as the one of 1966 would result in a water crisis scenario in the WMA (Sheer 1981). This specter catalyzed a sequence of events that ultimately resulted in a long-term solution to the WMA water supply problem. The solution had political, legal, social, and environmental dimensions—but technology played a major role.

Background

During the 1950s, the USACE undertook a number of studies of the Potomac River Basin, and in 1963 it recommended construction of 16 reservoirs for flood control and water supply (McGarry 1990). In 1969, the proposal was reviewed and it was recommended that six of the reservoirs were urgently needed and should be constructed as soon as possible. The Secretary of the Army concurred with the recommendation in 1970 but assigned high priority to the Sixes Bridge and Verona Reservoirs. Congress accepted the Secretary's recommendation in 1974 and authorized design of the two dams, but due to strong citizen opposition to dam construction at that time, it also directed construction of a pilot water treatment plant to evaluate the potential of treating water in the Potomac estuary as an alternative source of water supply. Congress also directed that there be another WMA water supply study. The result of all of the studies was authorization of only two reservoirs. Directing attention to the pilot water treatment plant and a new WMA water supply study eroded confidence that the two authorized reservoirs would ever be built.

In 1962, while the studies discussed above were underway, one of the 16 reservoirs originally recommended by the USACE, a large Potomac River Reservoir at Bloomington, Maryland, was authorized for water supply and flood control. The project was also part of the Appalachian Redevelopment Program intended to create jobs. There was no opposition to the project and construction began in 1975. Figure 1 shows the Potomac River Basin and the reservoirs that currently (2006) provide water supply to the Washington D. C. Metropolitan Area.

In the Water Resources Development Act of 1974, Congress directed that the USACE not issue any permits for future water withdrawals from the Potomac River until the WMA users had agreed on how they would allocate the water in the Potomac during periods of low flow (McGarry 1990). This mandate served notice that if the players did not want to accept the federal solution (dams), they would have to decide how to meet D.C.'s needs during drought or they would not be able to increase the number of intakes. This was a serious matter since both the WSSC and FCWA needed new intakes on the Potomac.

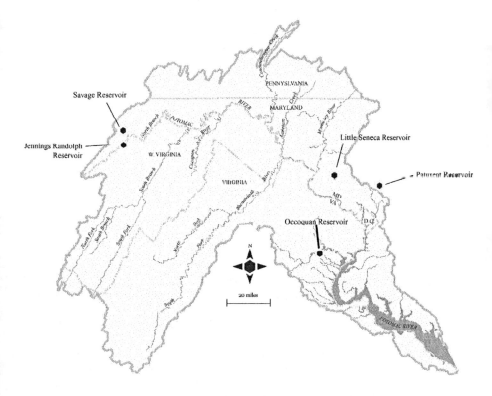

Figure 1. Potomac River Basin, basin states, regional water supply reservoirs.
Courtesy of the Interstate Commission on the Potomac River Basin, Rockville, MD

By 1976, it was generally accepted that the Sixes Bridge and Verona Reservoirs would not be built (McGarry 1990). It was recognized that the reservoirs would inundate large tracts of land and have major environmental impacts. Furthermore, the structural approach to water management was losing support. The public believed that greater emphasis should be placed on environmental protection and citizen involvement in planning processes; that local jurisdictions had not been appropriately involved in previous water supply studies; and that the quantity of water needed to supply the WMA had not been adequately addressed. Two questions were posed: Can the WMA get by with less water? If so, would the other dams be needed (McGarry 1990)?

Maryland Bi-County Water Supply Task Force

In 1975, dissatisfied with the stagnation of the federal process, elected officials of Prince George's and Montgomery Counties, Maryland and the WSSC formed a task force to see if a local solution to the two-county water supply problem could be found (McGarry 1990). The task force had strong support from local leaders who recognized the need to address the water supply problem. The task force recognized the importance of getting citizen and political consensus on solutions to the problem, and it embraced the notion of devising a technical solution based on accepting and managing droughts rather than on planning to provide all of the water needed, even during droughts (McGarry 1990).

The task force was co-chaired by the presidents of each of the two County Councils. These political leaders would be involved in the eventual approval of task force recommendations. A Citizen Advisory Committee (CAC) was selected from interested environmentally oriented citizens who would be involved in every step of the decision-making process. A technical group provided supporting engineering and environmental studies. Their work was directed by the task force with strong input from the CAC. The CAC insisted that solutions that might not be "fail safe" be developed (McGarry 1990). Technical support was provided by the WSSC and consultants.

The technical solution to the bi-county problem hinged on acceptance of the premise that droughts should be accepted and managed and that the plan should not be to meet maximum demands during critical low flow periods. Embracing the concept of drought management was an important feature of the technical solution. Studies showed that if the region would accept water use restrictions during severe droughts, the demand could be reduced significantly. The task force determined that the counties should accept an 8% risk that there would be limited shortages in any year and that water use would have to be restricted (McGarry 1990). Accepting this risk reduced the water supply needs by two-thirds, with significant reductions in cost and environmental impact. This concept was adopted. By taking this course of action and proposing a small dam on Little Seneca Creek, it was projected that bi-county water needs would be met through 2010.

Potomac River Low-Flow Agreement

The FCWA and the WSSC recognized that they would not be able to get the additional intakes they needed on the Potomac River without a low-flow agreement (McGarry 1990). As a result of negotiations with the USACE and the District of

16

Columbia, it was agreed that two intakes would be allowed, provided that the District would not be harmed by growth in the suburbs of Maryland and Virginia. It was agreed that the allocations of water to the FCWA and the WSSC during a drought be based on the winter demands of the three jurisdictions in 1975. For the District, it was determined that needs would be provided for, but for the WSSC and FCWA with their growing populations, this allocation would become less and less adequate over time. This condition imposed a significant constraint on the Maryland bi-county solution to its water supply problem.

Another impediment to the Maryland bi-county solution was the questionable likelihood of constructing the Little Seneca Dam without federal funds. Furthermore, the USACE and the EPA indicated that they would be reluctant to issue permits for construction of the dam without an accepted regional plan.

WMA Water Supply Study and Task Force

The first phase of the USACE WMA water supply study was completed in 1980. It showed that the ultimate solution to the water supply problem could be implemented by the three local government jurisdictions (McGarry 1990). On the strength of this pronouncement, it was decided that a WMA Water Supply Task Force be established with a format similar to that of the Maryland Bi-County Water Supply Task Force.

The task force was co-chaired by the presidents of each of the three County Councils and the District of Columbia. This political leadership group would, as before, be responsible for the approval of task force recommendations. A CAC was selected from interested environmentally oriented citizens who would be involved in every step of the decision-making process. A technical group, chaired by the general managers of the water supply agencies, took the lead in developing a new approach to water supply management. Technical support was provided by the Interstate Commission on the Potomac River Basin CO-OP (ICPRB's Section for Co-operative Water Supply Operations on the Potomac), WSSC, FCWA, WAD, and consultants. The technical staff developed supporting engineering and environmental studies.

The task force work plan included demand analysis (USACE MWA study used), existing capacity determination, identification of options for meeting shortage, public workshops, development of action plans, public hearings, and selection of an action plan (McGarry 1981).

All regional facilities were included in determining the capacity of the existing water supply system: Bloomington Reservoir (now Jennings Randolph, owned and operated by the USACE), WSSC's Patuxent River Reservoirs, the Savage Reservoir (owned by the Upper Potomac River Commission and operated by the USACE), WSSC's planned Little Seneca Reservoir, and FCWA's Occoquan Reservoir. The capacity studies, carried out by the ICPRB CO-OP, included daily demand, daily flows, and daily releases versus 0.38 million m^3/day (100 mgd) environmental flow-by (McGarry 1981).

The Role of Technology

In 1977, while working on problems associated with water quality management in the Potomac River Basin, Dr. Sheer of the ICPRB conjectured that withdrawals from the Potomac River by the utilities should be maximized continuously (Sheer 1981). This action would conserve water in the river system's reservoirs. By comparing demands, river flows, and storage, it was found that there was more than enough water in storage to meet water demands through the turn of the century. At the time, that was a surprising conclusion (Sheer 1981).

A 1977 drought in the Occoquan River Basin led to development of risk-analysis techniques for the Occoquan reservoir. In August 1977, the Occoquan Reservoir was being rapidly depleted and the problem was designated as serious (Sheer 1981). Analyses conducted by ICPRB indicated that the probability of falling below the "panic level" production rate of 0.15 million m^3/day (40 mgd) for the Occoquan Reservoir was about 13 percent (Sheer 1981). The local elected officials thought this was too high and wondered how much the demand would have to be reduced. By reducing the demand to 0.12 million m^3/day (32 mgd), it was found that the chance of falling below that production rate would be about 5 percent. The local governments then agreed to carry out a campaign to reduce the demand to 0.12 million m^3/day (32 mgd). This was the first time that risk analysis had been used in the WMA (Sheer 1981).

As a result of mounting public opposition to structural options for solving the WMA water supply problem, researchers at Johns Hopkins University and ICPRB began seeking alternative solutions. The resulting studies showed that coordinated use of the water stored in reservoirs in the Potomac River Basin during droughts largely eliminated the need for most new reservoirs (Hagen et al. 2005; Sheer 1981; Palmer et al. 1979).

The Hopkins research showed that by efficiently operating the WMA water supply system as a whole, the potential yield at Washington, D. C. would exceed 3.79 million m^3/day (one bgd) (Sheer 1981). Trade-offs between upstream reservoir release requirements and years beyond 1980 for meeting the water supply needs of the WMA are shown in Figure 2 (Palmer *et al.* 1979; Sheer 1981). The results are striking in that they illustrate that even over the range of required upstream reservoir releases shown, the water supply requirements of the WMA could be met until at least about 2025.

The ICPRB and Hopkins studies also proved that the WMA water supply system should be analyzed daily rather than monthly. Dr. Sheer's model showed that large releases from the Jennings Randolph Reservoir based on monthly projections would waste considerable water because the released flows by-passed Potomac River intakes during low daily demand periods (McGarry 1990).

Historically, the three water supply agencies had been operating their systems independently, concerned with what they considered best for their constituents. No consideration was given to adopting a regional perspective. The technical studies of the 1970s by the ICPRB and Hopkins researchers disclosed that if the three utilities were operated as a system and if releases from Jennings Randolph were made daily, only a small portion of the storage originally proposed by the USACE would be needed (Hagen et al. 2005).

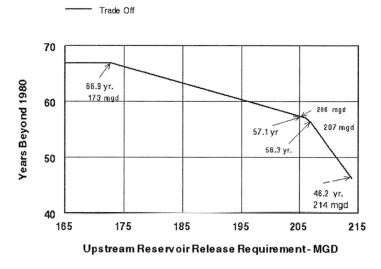

Figure 2. Trade-off curve for upstream reservoir release requirement and system yield
(Palmer et al.1979, Sheer 1981).

In their analyses, the researchers considered the seven-day travel time from Jennings Randolph to the WMA intakes. By managing Jennings Randolph collectively with the existing Occoquan and Patuxent River reservoirs it was found that there would be enough water for growth in the WMA through 2020, even with the occurrence of record droughts such as those of 1930 and 1966 (Hagen et al. 2005). It was also determined that system reliability would be achieved by adopting operating policies that required the WMA utilities to depend more on the free-flowing Potomac River during low flow winter and spring periods so as to preserve storage in the Patuxent and Occoquan River Reservoirs (Hagen et al. 2005). It was determined that this would work since, even during drought months, Potomac River flows exceed water supply demands. Such a policy reduces the risk of system failure and ensures availability of storage in the Occoquan and Patuxent Reservoirs to meet demands during summer low flows (Hagen et al. 2005). It was recognized that implementation of such an operating policy would require regional cooperation and the development of legal, financial, and operational agreements.

The water supply utilities and the USACE agreed that a simulation exercise to determine how the reservoir and river system would perform under various operating conditions would be a way to test the value of a coordinated operating policy for the Potomac (Sheer 1981). The initial model used was designated (PRISM), the Potomac River Interactive Simulation Model (Palmer et al. 1979; Hagen et al. 2006). Results obtained using PRISM were instrumental in bringing about consensus supporting the cooperative policy agreed to in the Water Supply Coordination Agreement (Hagen et al. 2006).

Since the late 1970s, the PRISM model has undergone several modifications and has been renamed PRRISM, the Potomac Reservoir and River Simulation Model (Hagen et al. 2006). The current version of PRRISM (2006) was developed for the demand and resource studies using the object-oriented programming language Extend™ (Hagen et al. 2006). PRRISM uses a water balance at the reservoirs and simulates river flows over a period of record. The model is used to evaluate the response of the current or modified system of reservoirs and the Potomac River to present or future water demands using current reservoir operating policies and the historical stream flow record (Hagen et al. 2006). The model is also used to play drought management games, an important means of evaluating or developing drought management policies. PRRISM is the primary tool for conducting the resource assessment studies that are scheduled on a five-year basis (Hagen et al. 2006). In a sense this may be considered an adaptive management approach to water supply planning and management in the WMA.

Implementation of the WMA Water Supply Management Plan

The WMA Task Force, having accepted the concept of cooperative water supply management, directed the Technical Group to work out the contractual agreements needed to support system operations and cost-sharing of facilities. This required execution of eight contracts involving three counties, two states, two independent water supply agencies, and the USACE (for the federal government).

On July 2, 1982, at a historic ceremony in the District of Columbia Building, the contracts were signed assuring the WMA of an adequate water supply until 2050 (McGarry 1990). It was noted that the regional water supply system would be completed for about $31 million, whereas the cost of the federal reservoirs that had been proposed would have been about $400 million (McGarry 1990).

Summary

In the late 1970s, the WMA political and institutional leadership recognized the value of exploring non-traditional technical alternatives for solving their water supply problems. As a result, the USACE, Maryland and Virginia, the District of Columbia, the ICPRB, the FCWA, the WSSC, the Metropolitan Washington Area Council of Governments, and other stakeholders began working together to provide a forum for coordinating their water-management policies.

A unique marriage of institutional cooperation and technical expertise resulted in a solution to the water supply problem of the WMA that had been sought unsuccessfully since the 1950s. A succession of studies had terminated with a recommendation that the only way to meet the future water supply needs of the WMA would be to construct two more very large reservoirs. But this belief was proven false when analysts at the ICPRB and Johns Hopkins University found that operating all existing reservoirs and utilities as a single coordinated system, something that never had been done before, would eliminate the need for additional reservoirs, at least until 2020 (Sheer 1981, McGarry 1990, Hagen et al. 2005).

The WMA water supply problem was solved because (1) leaders of the three water supply agencies were committed to finding a solution, (2) citizen leaders were involved from the outset and concurred in making decisions, (3) there was strong dedication by those in leadership roles to solve the water supply problem, and (4) traditional planning concepts were abandoned and replaced by innovative new water management approaches. The solution to the problem was based on cooperative systems management rather than on structural development and had the following principal elements (ASCE 1983):

- Using optimization and simulation models to develop practical rules for coordinated operation of the WMA water supply system.
- Integrating the National Weather Service River Forecast System (soil moisture accounting-based) with reservoir operations.
- Developing water demand forecasting models for projecting future water needs.
- Developing operating procedures for the complex WMA water distribution system based on system analyses and hydrologic models.
- Using risk analysis to identify the start of potential droughts and quantifying risks associated with drought.
- Using "drought games" to test and improve water supply operating policies and using these games as educational tools for decision makers.

The benefits of coordinated water management were found to be substantial in terms of meeting projected water demands, minimizing cost, and maximizing environmental protection (Sheer 1981, Viessman and Welty 1985). It was calculated that if all of the water supply facilities of the WMA were independently operated, the total yield of the system would be about 2.35 million m^3/day (620 mgd), but if the system were operated cooperatively, the yield would exceed 3.12 million m^3/day (825 mgd) (Sheer 1981). The resulting increase in yield exceeds 25 percent and is about equal to the yield associated with construction of the two reservoirs that had been under consideration (Sheer 1981). It is clear that technology played a considerable role in solving the WMA water supply problem.

21

References

Anon. (June, 1983). "Water supply," *Civil Engineering (ASCE)*, vol. 53, no. 6, Reston, VA.

Hagen, E. R., Holmes, K. J., Kiang, J. E., and Steiner, R. S. (December, 2005). "Benefits of iterative water supply forecasting in the Washington, D.C., metropolitan area," *J. of the American Water Resources Association (JAWRA)*, Middleburg, VA, 1417-1430.

McGarry, R. S. (Nov. 12, 1981). "Outlook for the future as projected by the Task Force for the Washington Metropolitan Area" in "*A 1980s view of water management in the Potomac River basin*," *Report of the Committee on Governmental Affairs*, U. S. Congress, Senate, 97th Congress, 2d Sess., U.S. Gov't. Print. Off., Washington, D.C.

McGarry, R. S. (1990). "Negotiating water supply management agreements for the National Capitol Region," in *managing water-related conflicts: the engineer's role, edited by Viessman, W. and Smerdon, E. T.,* ASCE, 116-130.

Palmer, R. N., Wright, J. R., Smith, J. A., Cohon, J. L., and ReVelle, C. S. (1979). "Volume I, policy analysis of reservoir operation in the Potomac River basin," "Executive summary," *Water Resources Research Center, report, B-027-MD, University of Maryland,* College Park, Maryland.

Palmer, R. N., Wright, J. R., Smith, J. A., Cohon, J. L., and ReVelle, C. S. (1979). "Volume II, policy analysis of reservoir operation in the Potomac River basin, "Washington metropolitan water supply system," *Water Resources Research Center, report, B-027-MD, University of Maryland,* College Park, Maryland.

Palmer, R. N., Wright, J. R., Smith, J. A., Cohon, J. L., and ReVelle, C. S. (1979). "Volume III, policy analysis of reservoir operation in the Potomac River basin, "Potomac River interactive simulation model," *University of Maryland, Water Resources Research Center, report, B-027-MD,* College Park, Maryland.

Sheer, D. P. (Nov. 12, 1981). "Assuring water supply for the Washington Metropolitan Area, -- twenty-five years of progress," in "*A 1980s view of water management in the Potomac River basin*," *Report of the Committee on Governmental Affairs,* U. S. Congress, Senate, 97th Congress, 2d Sess., U.S. Gov't. Print. Off., Washington, D.C.

Viessman, W., and Welty C. (1985). *Water management: technology and institutions*, Harper & Row, publishers, New York.

Wolman, A. (Nov. 12, 1981). "The Potomac River revisited for the Nth time," in "*A 1980s view of water management in the Potomac River basin*," *Report of the*

Committee on Governmental Affairs, U. S. Congress, Senate, 97th Congress, 2d Sess., U.S. Gov't. Print. Off., Washington, D.C.

Water Availability Modeling in Texas

Introduction

Texas has tremendously complex water resources. The second largest state in the U.S. both by area and population, it experiences tremendous variation in hydrology. Average annual rainfall totals vary from 203 mm (8.0 in) near El Paso in the west to 1600 mm (62.8 in) at Orange in the east (NCDC 2007). The state's surface waters occur in 23 major river basins, and groundwater resources include 9 major and 21 minor aquifers (TWDB 2006b). The state's large and diverse economy reached a gross state product of $982 billion in 2005 (TCPA 2007), which would make it the world's 10th largest economy if it were an independent nation. This economy used 27.3 km^3 (22.1 million acre-feet) of water in 2003. Agriculture accounts for 59% of state water use with many high-value crops such as Rio Grande Valley citrus and High Plains cotton completely dependent on irrigation (TWDB 2006a). The state's population of 22.9 million persons (2005 estimate) has grown by 35% since 1990. The cities of Houston, San Antonio, Dallas, Austin, Fort Worth, and El Paso are among the nation's 25 most populous cities, and San Antonio, Austin, Fort Worth, and El Paso are also among the 25 fastest growing cities in the U.S. (U.S. Census 2006). Average municipal water use for this population is 623 liters per person per day (165 gallons per person per day), partially driven by the state's warm climate. Texas industry includes a significant portion of the U.S. petroleum and chemical manufacturing sectors, and the state's electric power production is 97% from thermal sources requiring significant cooling water supplies. Most of these defining characteristics of Texas' economy and population have developed since the 1950s, with continuation of these trends expected into the future.

Water resources planning and management in Texas before the mid-1990s followed an unusual historical path. Due to the state's unique history, water rights and law were an amalgamation of Spanish, Mexican, Republic of Texas, and U.S./State of Texas constructs. A riparian rights system for surface water was followed before 1889. State legislation in that year and in 1895 established the prior appropriations doctrine for new rights but recognized existing riparian rights. The next seven decades witnessed an increasingly unworkable situation as repeated attempts to quantify riparian claims failed. This lack of information, coupled with increasing appropriation of surface waters for new claims and a major drought in the 1950s, led to recognition of the inherent incompatibility of the two legal doctrines. The Water Rights Adjudication Act of 1967 established prior appropriations as the sole legal doctrine for surface waters in Texas and required a process of adjudication to harmonize all existing rights into a single priority system. The adjudication process lasted until the late 1980s (Bowman 1993; Wurbs 1995).

Groundwater rights in Texas were established separately from surface water rights by a 1904 Texas Supreme Court decision. That case established the "rule of capture" for groundwater wherein a landowner may "capture" from a well on the landowner's property any quantity of water he or she wishes, regardless of any injury that may occur to neighboring landowners. The court's decision was driven primarily by a recognition of the contemporary lack of knowledge on groundwater quantities and flow, described by the court as being "secret, occult and concealed" (Potter 2004). The rule of capture remained largely untouched until the mid-1990s.

Because the legal regime focused on individual property rights for groundwater, collective planning and management of groundwater resources was almost non-existent before the mid-1990s. In contrast, surface water planning and management were carried out in a centralized fashion effectively by a single state government agency that was often distant from local stakeholders. Surface water planning was mandated by the Texas Legislature after a severe drought in the 1950s. State water plans were published in 1961, 1968, 1984, 1990, 1992, and 1997, by the Texas Water Development Board (TWDB) (by its predecessor agency in 1961). As its name implies, TWDB's mission is to provide planning, technical assistance, and financial resources to Texas communities' efforts to meet water supply needs. However, other water-relevant functions are carried out by agencies such as the Texas Commission on Environmental Quality (TCEQ), which enforces environmental regulations and manages surface water rights, and the Texas Parks and Wildlife Department (TPWD), which is charged with maintaining healthy aquatic ecosystems. TCEQ and TPWD were not involved in the state water planning process until the 1992 plan. Local and regional stakeholders' input was usually limited to comments by outside advisory panels. This isolated planning process led to plans that had, by TWDB's own admission, "a limited power of persuasion in guiding the State's water future" (TWDB 1997).

The Drought of 1995-1998 and Senate Bill 1

The worst recorded drought in Texas history occurred from 1950 to 1957, early in the state's modern development. While this drought was devastating to the state's agricultural industry and brought municipal water supplies to low levels, it was followed by a relatively wet period lasting for decades. As Texas' population and economy boomed, droughts that did occur were relatively moderate in scope and duration. This good fortune ended with a drought that began in late 1995 and quickly strained the state's water supplies. That drought's effects included over $11 billion in agricultural losses, a drop in statewide reservoir levels to 68 percent of conservation storage, the implementation of demand management measures by more than 300 cities and water utilities, almost 500,000 acres burned by wildfires, and more than 14,000 farm workers out of jobs (TWDB 2007).

The Texas Legislature responded to the drought crisis in its 1997 session by passing the Brown-Lewis Water Management Plan, a.k.a. "Senate Bill 1" (SB1), the number reserved for the most important legislation considered in each session. SB1 called for wholesale change in water resources planning, management, and development in Texas. A new system of regional water plans was established with mandatory inclusion of representatives from a wide set of interests. As shown in Figure 3, the state was divided into 16 water planning regions based primarily on watershed boundaries. Regional planning groups were formed as committees composed of members representing, at minimum, municipalities, counties, industry, agriculture, environmental interests, small businesses, electric generating utilities, river authorities, water districts, water utilities, and the general public. The regional planning groups were to be provided resources and technical assistance through TWDB to prepare a 50-year plan, and, once interregional conflicts were addressed, the statewide water plan was to be composed as the collective of the regional plans. The planning process was to be repeated every five years. Thus, the old "top-down" approach was replaced by a "bottom-up" one.

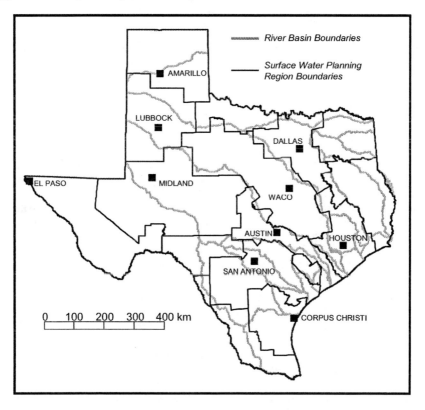

Figure 3. Texas surface water planning regions established by SB1 and major river basins.

SB1 required a change in water management philosophy, as the 1995-1998 drought had exposed the severe over-allocation of some water sources. Conservation was a mandated water management strategy in all regional plans. All new surface water right applicants would be required to include water conservation plans in their permit applications, and existing permit holders for amounts above certain levels would be required to file conservation plans.

Finally, SB1 recognized the lack of knowledge on what water supplies remained in the state for further development. "Water availability models" were mandated for all of the state's river basins for three stated purposes: First, to inform all water right holders in each basin what water remained available for development under various drought conditions. Second, to provide water availability data to the surface water regional planning groups. Third, to determine "the potential impact of reusing municipal and industrial effluent on existing water rights, instream uses, and freshwater inflows to bays and estuaries" (Texas Legislature 2006).

The call for water availability models by SB1 could not have come at a more fortuitous time. While the stated purposes of the models were limited, other water management issues in Texas at the time—principally the need for better tools with which to manage thousands of surface water rights permits as well as new applications—required much of the same technology as the availability models. Several technologies were also maturing at the time that greatly expanded the capacity of water resources practitioners to solve the state's needs.

The Texas Water Availability Models

The development of the Water Availability Models (WAMs) for each of Texas' river basins was actually composed of three primary technological steps: (1) development of naturalized streamflow databases, (2) spatial analysis using geographic information systems (GIS) in support of streamflow naturalization, and (3) water balance modeling accounting for all permitted surface water withdrawals and impoundments. A key aspect of the technology development process was its collaborative nature. The four-year process of model development involved multiple partners from state government (TWDB, TPWD, Texas Natural Resources Conservation Commission [TNRCC, the predecessor to TCEQ], Texas Agricultural Experiment Station), federal government (U.S. Geological Survey [USGS], Agricultural Research Service), academics (Texas A&M University, University of Texas), and several private engineering consulting firms. Extensive peer review was used throughout the process resulting in a general consensus on the high quality of the technology used and widespread understanding of how the technology works (the review process is discussed in more detail by Wurbs and Sisson 1999 and TNRCC 1999).

The first major task, determination of naturalized streamflows, was necessary to establish a baseline of surface water availability in the absence of human activity. As described above, people have extracted and impounded surface waters throughout Texas since before reliable stream gauging was used. Moreover, the scope of human impact has continuously changed over the past century. Thus, the full record of measured streamflows throughout the state includes multiple anthropogenic effects (i.e., diminishment by withdrawals and reservoir evaporation and augmentation by various discharges). In addition to the problem of these human-caused alterations is a problem of data scarcity, an inadequate number of operating stream gauges throughout all river lengths and historical times. One of the legislated functions of the WAMs is to provide water supply reliability information to all surface water permit holders in each basin. As an example, the Brazos River basin includes well over 1,000 permit holders in a 118,000 km^2 (45,600 mi^2) area (Wurbs 2001); however, the USGS currently operates only 62 stream gauges in this basin. There have been only 137 gauges to ever operate in the Brazos basin, and their periods of operation vary a great deal (USGS 2007). Thus, some procedure was needed to estimate naturalized flows at all permitted withdrawal and impoundment locations for all months in the period of record to be used for reliability analysis. The technical methodology for streamflow naturalization and estimation at ungauged sites is described by Wurbs and Sisson (1999). A brief summary is given here.

First, flow naturalization is accomplished for monthly flow volumes at gauged points using:

$$Q_{nat} = Q_g - W - R + E - \Delta S \qquad (1)$$

where Q_{nat} and Q_g = naturalized and gauged streamflows at a point, respectively, W = the sum of all withdrawals above the point, R = the sum of all return flows, E = reservoir evaporation, and ΔS = the change in storage of upstream reservoirs. This calculation obviously relies on records and/or estimates of several types and from several sources. The beginning record date for most WAM analysis was chosen to be January 1940, with continual updating of the naturalized record to the present. Thus, the state's drought of record of the mid-1950s and the state's most economically damaging drought of the 1990s are both included.

After all gauged monthly flow volumes have been naturalized, values for ungauged locations are estimated using:

$$Q_{nat,ungauged} = C \, Q_{nat,gauged} \qquad (2)$$

$$C = \left(\frac{A_{ungauged}}{A_{gauged}} \right) \left(\frac{CN_{ungauged}}{CN_{gauged}} \right) \left(\frac{M_{ungauged}}{M_{gauged}} \right) \qquad (3)$$

where Q_{nat} = naturalized monthly flow at ungauged and gauged sites as indicated, C = a scaling coefficient, A = total drainage area above ungauged and gauged sites as indicated, CN = the Natural Resources Conservation Service (NRCS) runoff curve number for the ungauged and gauged sites, and M = mean monthly rainfall for the ungauged and gauged basins.

The form of the streamflow naturalization equations requires significant spatial information for the thousands of ungauged basins created by Texas' distributed surface water permits. The need for this spatial information drove the second major technological task: spatial analysis of the state's river basins using GIS. This work is discussed at length by Hudgens and Maidment (1999). During the 1990s researchers at the University of Texas Center for Research in Water Resources, led by Drs. David Maidment and Francisco Olivera, pioneered the application of GIS to hydrologic applications. The first widely available GIS-hydrologic tool CRWR-PrePro was published in 1998 (Olivera et al. 1998), and continues in use today as HEC-GeoHMS.

The new technology possessed the capabilities to automatically delineate watersheds from raster digital elevation model (DEM) data and to perform spatial analyses of areas covered by soils, land cover types, and rainfall data. These traits made it perfectly suited to the data needs of the streamflow naturalization methodology. While the use of GIS has since become commonplace in water resources planning and management, the application of GIS analysis to determine streamflow naturalization parameters for the over 8,000 surface water permits and 696,000 km^2 (269,000 mi^2) of land area in Texas was perhaps one of the very first large-scale applications of GIS to water resources planning.

The final part of the WAM effort was development of a water balance model for each river basin that could calculate streamflow levels at all control points of interest (i.e., gauges and permit locations) from the naturalized streamflows net permitted withdrawals, impoundments, and other human-made factors. A comparative evaluation of 19 available water resources models resulted in the selection of the Water Rights Analysis Package (WRAP) (TNRCC 1999). WRAP had been under development since the mid-1980s by Dr. Ralph Wurbs at Texas A&M University as a prototype computer model. While WRAP had been used for several research applications, it had not been developed for full public use before SB1. The original impetus for this model was the completion of Texas' adjudication of surface water rights and the perceived need for an analytical tool for management of the thousands of prior appropriations rights that resulted (Wurbs 2005b). The 10 years of development invested in the model before SB1 had resulted in a highly refined understanding of reliability issues for surface water supplies, which made WRAP perfectly suited for the WAM water balance need.

WRAP is fully described by Wurbs (2001, 2005a). The model works on a monthly time step and simulates streamflow at all control points of interest in a river basin as the net sum of naturalized streamflow, upstream withdrawals, return flows, changes in reservoir storage, and reservoir evaporation. Withdrawals, returns, and storage changes are determined according to surface water permit characteristics. Model runs typically are made for the full period of record from January 1940 to the present. Output includes various measures of water supply reliability, among many other quantities. Examples of these reliability measures include "volume reliability" R_V and "period reliability" R_P:

$$R_V = \frac{v}{V}(100\%) \qquad (4)$$

$$R_P = \frac{n}{N}(100\%) \qquad (5)$$

where v = the total volume of water supplied to a permit holder over the simulation period, V = the total volume demanded over the simulation period, n = the number of months in the simulation where demand is fully met, and N = the total number of months in the simulation. Thus, WRAP brought to the WAM system the ability to quantify reliability for hypothetical water infrastructure projects and new water uses as measured against historical hydrology but under contemporary surface water appropriations.

While much of the initial technology development was accomplished by researchers at Texas universities, much of the data processing, quality control, and systems analysis were performed by several private consulting firms. In the end, the development of the Texas WAM in response to SB1 lasted from May 1997 to December 2001 and cost approximately $4.7 million (TNRCC 1999).

How the Texas WAMs Are Used

The Texas Water Availability Models have become a tremendously important technology for planning and management of surface water resources in the state. As was discussed above, one of the legislated purposes for the models was to provide data to inform the regional planning process performed by the 16 surface water planning groups. The WAMs have indeed been capable of producing reliable information on remaining surface water resources for future development. Because of the open and technically sound process of model and database development and the common modeling framework for all regional groups, the WAMs allow efficient coordination between groups that share common water resources with very few disagreements on planning data. Model results are viewed with high confidence, and disagreements between regions remain in the political sphere. The richness of the information provided by the WAMs also allows for very sophisticated planning exercises. A crude measure of how water planning has grown with better data is the length of final planning documents. The 1997 State Water Plan (the last completed under before the WAMs and the "bottom-up" approach) was 340 pages long, excluding appendices. The 2007 Region C Plan (which includes the Dallas/Fort Worth area) alone is over 485 pages long, excluding appendices. The appendix to the 2007 Region C Plan that presents water availability data determined by WAM simulations is 41 pages long.

The emphasis on reliability analysis built into the WAMs has also enriched the management and regulatory processes. TCEQ, the state agency charged with evaluating new surface water right applications, has established criteria that municipal supply permits must have 100% volume and period reliabilities and agricultural permits must have 75% volume and period reliabilities (Wurbs 2005b).

Thus, the WAM systems are now used to prepare and evaluate new surface water permits.

In addition to the various agencies and groups discussed above, Texas also has 13 river basin authorities chartered by the state with various basin management activities, such as wholesale water supply, hydropower production, flood control, community development, and reservoir operation, among others. The capacities of the WAMs to support decisions have naturally allowed these river basin authorities to change their own planning and management processes. One example of this change is the Brazos River Authority (BRA), which operates a system of 11 reservoirs on the Brazos River and its tributaries. In 2005, BRA applied to the state for a "systems operation" permit, a new form of surface water right that recognizes how water can be made newly "available" by coordinated operation of a system of reservoirs. The BRA application claimed additional supply capabilities of over 0.49 km^3 (400,000 acre-feet) per year under system operations as modeled by the WAM (Brazos G Regional Water Planning Group 2006).

Lessons Learned and Conclusions

The Texas WAM effort has proven to be remarkably successful. The technology, in concert with significant institutional change, has changed a relatively ineffectual state water planning process into one of the most sophisticated and vibrant in the U.S. It is used daily by a wide range of users in industry, government, and academia for a full slate of planning, regulatory, management, and research applications.

Several factors that are key to the success of the WAM project can be lessons for other water resources planning and management technologies. First, collaborative development was essential. The WAM development process included significant involvement by multiple members of the academic, government, and industry communities. Major decisions on technological and methodological issues were submitted to a transparent and publicly documented peer-review process. As a result, there was wide understanding of and confidence in the technologies used in the final system and why they had been chosen. The involvement of multiple state government agencies allowed development of tools that could be used for multiple functions; thus, a single system is now used for both long-term regional and statewide planning (overseen by TWDB) and for evaluation of individual surface water permits (overseen by TCEQ). Inclusion of private consultants in the system development process ensured proficiency with the system when their services were immediately needed for the 2001 regional planning cycle. Inclusion of the state's universities not only capitalized on their technology generation capacity, but also formed the academic route whereby civil engineering students are now trained in the use of the Texas WAM system before entering the professional workforce.

Second, the Texas WAM system has remained in a continual improvement mode since its inception. WRAP, the computational heart of the model, was itself a model that had been developing for a decade when it was chosen for inclusion in the WAM.

Like the example of system operation permitting described above, Texas water resources management is continually in a dynamic process of re-invention, and the WAM software is being updated to address these changes. Ongoing problems such as naturally occurring salinity and new issues such as instream flows, natural channel conveyance of reuse water, conditional reliability, and the need for daily management simulation have been incorporated in the WAM over time.

The Texas Water Availability Models have been a tremendous asset to water resources planning and management. In a decade they have transformed the process by which Texas understands and manages its water future.

References

Bowman, J.A. (1993). "Reallocating Texas' water: Slicing up the leftover pie," *Texas Water Resources,* Texas Water Resources Institute, 19(4), 1-11.

Brazos G Regional Water Planning Group. (2006). *Regional Plan,* Texas Water Development Board, Austin, TX.

Hudgens, B.T., and Maidment, D.R. (1999). "Geospatial data in water availability modeling." *CRWR Online Report 99-4, Center for Research in Water Resources, University of Texas,* Austin, TX.

National Climatic Data Center (NCDC), "U.S. Climate Normals," http://cdo.ncdc.noaa.gov/cgi-bin/climatenormals/climatenormals.pl. (Jan. 3, 2007).

Olivera, F., Reed, S., and Maidment, D. (1998). "HEC-PrePro v. 2.0: An ArcView pre-processor for HEC's Hydrologic Modeling System," in *1998 ESRI User's Conference, July 25-31, 1998,* ESRI, Redlands, CA.

Potter, H.G. (2004). "Chapter 1: History and evolution of the rule of capture," in *Report 361 Conference Proceedings, 100 Years of Rule of Capture: From East to Groundwater Management, edited by Mullican, W.F., and Schwartz, S.,* Texas Water Development Board, Austin, TX, 1-10.

Texas Comptroller of Public Accounts (TCPA), "Economic Indicators," http://www.window.state.tx.us/ecodata/ecoind/ecoind.html. (Jan. 4, 2007).

Texas Legislature, "75(R) SB 1 Enrolled Version," http://www.capitol.state.tx.us/tlodocs/75R/billtext/html/SB00001F.htm. (Dec. 19, 2006).

Texas Natural Resources Conservation Commission (TNRCC), (1999). "WAM: Water availability modeling, an overview," *Publication GI-245, TNRCC,* Austin, TX.

Texas Water Development Board (TWDB), "2003 Water Use Survey Summary Estimates," http://www.twdb.state.tx.us/data/popwaterdemand/2003Projections/HistoricalWaterUse/2003WaterUse/HTML/2003state.htm. (Dec. 20, 2006a).

Texas Water Development Board (TWDB), "Mapping," http://www.twdb.state.tx.us/mapping/index.asp. (Dec. 20, 2006b).

Texas Water Development Board (TWDB), "The drought in perspective, 1996-1998," http://www.twdb.state.tx.us/data/drought/DroughtinPerspective.asp. (Jan. 8, 2007).

Texas Water Development Board (TWDB). 1997. "Water for Texas: A consensus-based update to the State Water Plan," *Document No. GP-6-2. TWDB,* Austin, TX.

U.S. Census Bureau, "U.S. Census Press Releases," http://www.census.gov/Press-Release/www/releases/archives/population/007001.html. (June 21, 2006).

U.S. Geological Survey (USGS). "USGS surface-water data for Texas," http://waterdata.usgs.gov/tx/nwis/sw. (Jan. 4, 2007).

Wurbs, R. (1995). "Water rights in Texas," *J. of Water Resources Planning and Management,* ASCE, 121(6), 447-454.

Wurbs, R. (2001). "Assessing water availability under a water rights priority system," *J. of Water Resources Planning and Management,* ASCE, 127(4), 235-243.

Wurbs, R.A. (2005a). "Fundamentals of water availability modeling with WRAP," TR-283, *Texas Water Resources Institute, Texas A&M University,* College Station, TX.

Wurbs, R. (2005b). "Texas water availability modeling system," *J. of Water Resources Planning and Management,* ASCE, 131(4), 270-279.

Wurbs, R., and Sisson, E. (1999). "Comparative evaluation of methods for distributing naturalized streamflows from gauged to ungauged sites," *TR-179, Texas Water Resources Institute, Texas A&M University,* College Station, TX.

Libya's Groundwater Development Project

Introduction

In 1953 the search for new oilfields in the deserts of southern Libya led to the discovery not only of oil but also of underground aquifers containing vast quantities of fresh water. Most of this water had been trapped there for more than 30 thousand years. The amount of this 'fossil' water is substantial, equivalent to a pool of water over 100 meters deep covering an area the size of Germany. This was good news for Libyans. Libya is located in one of the driest regions of the world. Its annual rainfall ranges from 10 to 500 millimeters. Only 5 percent of Libya receives more than 100 mm of rain annually. Further, evaporation rates are orders of magnitude higher than precipitation rates. Hence, like many other Middle Eastern countries Libya has been mining groundwater.

In 1974 Libyan planners began thinking about how these newly discovered water supplies could be exploited. Nine years later the Libyan government established the Great Man-Made River Authority to implement and manage the so-called Great Man-Made River Project (GMRP). This project was to be the greatest feat of civil engineering ever undertaken by mankind at that time – to bring this ocean of fossil water to the people of Libya. The project consists of manufacturing and then installing well over 1000 wells and pumps and associated infrastructure at multiple well fields, and over 5000 km of pipes of up to 4 meters in diameter to convey the groundwater from its sources under the Sahara desert to where it is needed in the Mediterranean coastal rim of the country.

Figure 4 illustrates the extent of this groundwater recovery project. The Great Man-Made River Project has been recorded in the Guinness Book of Records (1993) with respect to its cost, period of construction, number of personnel involved, and the special equipment and the technology used.

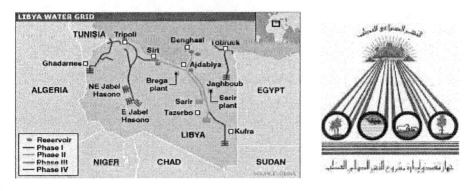

Figure 4. The Great Man-Made River Project in Libya (GMRA 2006).

Project Description

Libya covers an area of about 1.8 million square kilometers. The country spans three climatic zones: the Mediterranean, the semi-desert, and the vast desert zone of the northern Sahara. The present population of about five million lives mainly in the Mediterranean coastal zone, with a large proportion in its cities of Tripoli and Benghazi.

The project has proceeded in planned phases. The first and largest phase was completed in November 1994. The Phase 1 system is designed to supply two million cubic meters of water per day from well fields at Sarir and Tazerbo to end reservoirs at Sirt and Benghazi, the second-largest city in Libya. The depth of the wells exceeds 500 meters. Special pipe manufacturing plants were built at Sarir and Brega. Using groundwater from local wells the plants produced 7- to 8-meter pre-stressed pipe sections (weighing some 80 tons) that were then transported, via specially built carrier trucks, over a series of newly and specially built haulage roads to where they were needed (Figure 5). Once the pipes reached their destination, cranes and bulldozers placed them into trenches (Figure 6). The pipes were then connected, tested, and the trenches back filled. All this was done in temperatures reaching 50°C. At strategic locations pumps and surface water reservoirs were installed to handle particular hydraulic issues associated with the pressurized flow in the pipes. This engineering feat continues today, as the project is not yet complete.

Figure 5. Specialized trucks transporting pipe sections, day and night, into the desert.

Figure 6. Specialized equipment for digging and placing and
connecting pipe sections in the pipeline trenches.

Phase II of the Great Man Made River Project conveys two million cubic meters of water from well fields at East Jabal Hasouna and North East Jabal Hasaouna to Tarhouna and Tripoli, as shown in Figure 4. This phase includes 440 wells and 895 kilometers of 4.0- and 3.6-meter-diameter pipes with pressure ratings varying from six to 26 bars. Conveyance pump stations are located at each of the wellfields to convey the water downstream to a regulating reservoir. The water for Tripoli now comes from this system rather than from a depleted and increasingly saline coastal aquifer.

Phase III of the GMRP is the Gardabiya–Assdada link that connects the Phase I system at Sirt to the Branch of the Phase II system located at Assdada. The link has been designed to convey about a million cubic meters per day in either direction. This is achieved through two additional pump stations, one at Sirt and the other at Assdada.

The aim of Phase IV (which is sometimes divided into three separate phases) is to convey 90 million cubic meters per year from Ghedames basin to the costal regions of western Jefara area to meet the domestic water requirements of the growing population in that area. The wellfield consists of 144 wells with an additional 15 standby wells. They are drilled between 900 to 1100 meters deep. Conveyance pump stations are required.

Another part of Phase IV, sometimes referred to as Phase V, consists of wells and pipelines intended to pump and convey water from the Jaghboub area to the coastal regions which extend from the west Al Bamba Gulf to east Emsaad, supplying the Batnan population with water and also some of the east Al Goba regions.

The third part of Phase IV is the Kufra-Tazerbo conveyance that will involve developing a new major wellfield at Kufra and will increase the capacity of the Phase I System to 3.68 MCMD.

All of these phases, as illustrated in Figure 4, require the technology shown in Figures 5 and 6. What is not shown in these figures is the additional technology that includes the needed power generation and transmission systems, the communication and electronic remote control systems, and all of the maintenance facilities. This GMRP is indeed a technology driven project of the first order.

Some statistics just for Phases I and II may be of interest. Approximately 500,000 pre-stressed concrete cylinder pipes were manufactured, transported (Figure 5), and used. The distance traveled by the transporters was equivalent to a trip to the sun and back. Over 3,700 km of haul roads were constructed alongside the pipe line trench to enable the heavy trucks to deliver pipe to the installation site. The volume of trench excavation exceeded 250 million cubic meters. The amount of aggregate used in the project exceeded 30 million tons. The total weight of cement used exceeded 7 million tons. A road from Tripoli to Bombay could be paved with the cement used to build the pipes. The total length of pre-stressing steel wire in those pipes was about six million kilometers, enough to circle the earth 280 times.

If the conveyance system were placed on Western Europe, the pipeline would begin in southern Switzerland and go through Germany and on up to Poland before cutting west to northern Scotland. On average it takes about nine days for a drop of water to make that trip, excluding the time spent in storage tanks or reservoirs.

Water from the Great Man Made River Project will be used for agricultural (70%), industrial (2%), and domestic (28%) uses. Traditional water resources such as the coastal aquifers are becoming saline due to over use, risking collapse of agricultural lands. The project will permit a reduction in the extraction of water from Libya's coastal aquifers, as agriculture ceases to be dependent on existing water wells. The water that flows from the desert aquifers will also serve the Brega and Ras Lanuf industrial regions.

At planned extraction rates the groundwater supplies are expected to last at least 50 years. During this time, the coastal aquifers are expected to be replenished and Libya will increase its degree of food and water self-sufficiency.

Libya could supplement its water supplies in other ways. Bringing water to the country by pipe or by ships from foreign sources is possible technically, but perhaps difficult politically. Desalination of brackish and Mediterranean Sea water is also technically possible. During the planning of the GMRP these options were considered and their costs were estimated. The conclusion was that the cost of pumping and transporting Libya's underground water from the desert was more economical than any other alternative. At a total project cost of about $30 billion dollars, the unit price per cubic meter of fossil water at design capacities is less than a quarter of the cost of desalination.

After the completion of most of Phase I and II, with improved cost estimates available, the GMRA decided to conduct a cost-effectiveness analysis considering the remaining phases of the GMRP and desalination, its nearest competitor. The remainder of this case study describes this study that involved the use of optimization models to examine and estimate the costs of many combinations of project and desalination systems.

Cost-Effectiveness Analyses

In the mid 1990s the GMRA sought assistance from UNESCO to conduct a study to determine what combinations of groundwater and desalinated water would minimize the cost of meeting projected increasing demands over time at major demand sites in the coastal regions of Libya. The initial idea was to conceive a dozen or so combinations of groundwater and/or desalination infrastructure components and configurations and then determine the cost of each of these scenarios to identify some least-cost systems. The questions that needed to be answered included the following:

- How many wells in each potential wellfield should be installed—taking into account failures and maintenance requirements as well as water supply demands?
- What size pipes and combinations of pumps and surface reservoirs should be installed between the wellfields and demand sites to convey the water pumped from the wellfields to the demand sites?
- Which wellfields should serve which demand sites, if any? In other words, what system configurations were cost effective?

- Where and how much desalination capacity should be installed and what demand sites should they serve?
- What degree of redundancy or excess capacity is needed to achieve given levels of reliability in case of system component failures?
- How sensitive are the least-cost configurations to cost uncertainties, currency exchange, and interest rates?
- How can desired water quality (salinity) concentrations be met by mixing waters taken from different wellfields?

Costs were to include installation and annual operating and maintenance costs. Each project phase not yet completed was to be analyzed separately and then combinations of phases were to be analyzed.

Modeling Approach

After some discussions with Libyan GMRA engineers and directors, it seemed appropriate to develop a set of optimization models that would allow the consideration of a large number of discrete system component capacities for any particular configuration scenario. Each model was based on mass balances in node-link networks representing wellfield and desalination sites, conveyance pipelines, pumps and reservoirs, and demand sites. The unknown variables of the models were the flows through all the system components and the capacity of those components.

The solution of each model identified the minimum total annual cost of meeting any specified set of flow demands at various demand sites. These total costs were based on the cost functions of each system component. Costs depended on component capacities, and these were determined by the flows through those components. Model constraints were continuity of flows in all components and requirements that component capacities be no less than the flows through them. In other words, the greater the flows, the greater were the needed flow capacities and hence the greater were the costs. The optimization procedure would find the least-cost combination of component capacities for each system configuration and demand scenario.

Using this modeling technology, over 250 separate system configurations and demand scenarios were analyzed, not just a dozen as originally envisioned.

Obtaining Data

Obtaining the needed cost functions and hydraulic flow capacities of pipelines, pumps, and wells was a challenge, especially for systems not yet sufficiently designed by the design engineers. This is understandable. In cases where there was some uncertainty, ranges of likely costs were used, and the sensitivity of these cost uncertainties on the system design configuration and design capacities was determined.

The cost and other design data needed to perform these analyses had to be obtained from different governmental agencies. Although this was being done at the request of the Libyan government, this did not guarantee easy access to information from all applicable government agencies or even from the consulting design firm. Data is indeed power, and power is not always easily relinquished. Rarely do these aspects of model building and analysis in the real world get discussed in the professional literature.

Static and Dynamic Analyses

The first portion of the analyses did not consider time. The analyses were solved for fixed demands. They were designed to find the least-cost way to meet a specified set of fixed demands at specified demand sites. What became obvious is that the least-cost way to meet various fixed demands can result in a change in system configuration as well as in component capacities. Consider wellfields for example. Each wellfield has fixed development costs as well as variable capacity costs. Different wellfields have different maximum possible flow capacities. Figure 7 illustrates a possible simple example of such a situation.

As demands increase over time, the least-cost way to meet any specified demand may switch from a wellfield that has a relative low fixed development cost and relatively low maximum pumping capacity (Alternative B in Figure 7) to another wellfield (Alternative A in Figure 7) that has a higher fixed development cost but also a greater maximum pumping capacity. The lower cost and more limited capacity wellfield drops out of the solution and the wellfield with the higher cost becomes the least-cost one. Of course this makes no sense in reality if the lower cost wellfield was developed to meet lower demands. In this case its cost is still there when the higher cost wellfield is needed at a later time. Of course, both wellfields can be used to meet the total demand but this is more expensive than just developing the higher cost one.

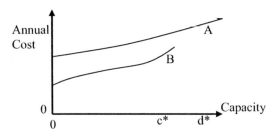

Figure 7. Fixed and variable costs of two alternatives for meeting capacity demands.

To meet a capacity demand of c^*, Alternative B is cheaper. Alternative A is not implemented and thus costs nothing. Its fixed costs occur only when it is implemented. Similarly, for a capacity demand of d^*, implementing Alternative A alone will be cheaper than implementing any feasible combination of both alternatives.

38

Static analyses cannot answer the question of which wellfield, e.g., Alternatives A or B in Figure 7, should be developed first as the demands increase over time from 0 to c* to d*. Thus a set of dynamic models was developed and used to identify the sequencing of component capacities that minimize the present value of total annual costs to meet increasing flow demand scenarios. These dynamic models constrained the capacities of components to only increase over time. Their costs continued over time once built. Costs of components, such as Alternative B in Figure 7, that would not be built for high-demand scenarios, such as for d* in Figure 7, but could be built for low-demand scenarios, such as to meet a demand for c*, remain as a cost even if the demand increases, such as from c* to d*. Thus the optimization determined whether or not a low-cost but low-capacity alternative should be implemented assuming that the demand will increase beyond its capacity over time. These dynamic models also allowed the consideration of component depreciation and component replacement.

Model Results and Use

The specifics of these analyses are not important for the purposes of this case study (for example see El Geriani, et al. 1998), but their results confirmed that desalination was more expensive than the GMRP over a range of interest and Libyan dinar-US dollar exchange rates, and uncertain component costs. However desalination was viewed as a way to increase system reliability in case of GMRP component failures.

Based on today's costs, the GMRA reports the total quantity of water that can be obtained per Libyan dinar (that will depend on the magnitude of water delivered, of course) is as shown in Figure 8.

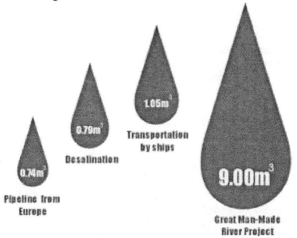

Figure 8. Quantities of water that can be obtained from various sources for a Libyan dinar (about 0.78 US$). (GMRA 2006)

The results of these models were submitted in reports to UNESCO and the GMRA (Loucks and Pallas 1997). In recognition that these model results alone would perhaps be less useful than the ability of GMRA personnel to modify the models and their input data and examine additional scenarios, workshops were held in Libya on the structure and use of these optimization models. The solution software these optimization models depended on was purchased and made available to those attending the workshops.

It is not clear just how much this study and its modeling technology changed the debate about the further development of the GMRP. Since this was an 'outside' study—but one that was requested by and involved and required the cooperation of GMRA personnel—confirmed the belief that the GMRP was more cost effective than its next feasible alternative, desalination, the GMRP has continued and even expanded beyond what was considered when the study was undertaken some 10 years ago.

Clearly the capacity being installed in the GMRP today exceeds the current and near-term demands as one would expect given the fixed costs. Nevertheless the extent of this excess installed capacity does not seem to be based on any sequencing and scheduling analysis. The capacity being installed for each system phase appears to be independent of the projected demands over time. The sequencing of the phases depends as much on the availability of financial and construction resources as it does on demand projections. The GMRA wants to build all phases of the entire project up to their 'design' capacities as soon as possible while they have the political support to do it. Some day in the future, such a strategy may indeed seem to have been an 'optimal' one.

Sustainability

Clearly, mining fossil water is not sustainable in the long run. Should this be a concern, as some might suggest? Indeed it should, just as mining a finite supply of coal, oil, and gold should be. This section concludes with some thoughts on the issue of sustainability.

Eventually this enormous supply of fossil water will diminish and become too expensive to extract compared to other alternatives. Until that happen, however, it is not a question of whether or not this non-renewable water should be pumped, transported, and used. It is rather a question of how much should be pumped, transported, and used over time and for what purposes. (Should water in the form of tomatoes grown from Libya's desert using fossil water be shipped to a much more water-rich Europe, for example?)

If we knew how, we could estimate the quantities of water to extract and use each year by equating the present values of each future year's marginal net social welfare benefits obtained from that water (Loucks 1994). This also assumes we know the right interest rate to use as well, and in Libya the preferred interest rate is 0! – but that is another story. In the absence of such data Libya should extract what it believes it can effectively and efficiently use each year to increase its standard of living and its food and water security and improve its physical and social infrastructure.

Perhaps by the time this fossil water is considered no longer a viable source, engineering technology and the political options available to future Libyans will have improved sufficiently to make other ways of meeting Libya's water demands attractive. While fossil water itself may not be sustainable, its wise use over the time it is available may increase the degree Libya itself becomes more sustainable.

References

El Geriani, A.M., Essamin, O., Gijsbers, P.J.A., and Loucks, D.P. (1998). "Cost-Effectiveness Analyses of Libya's Water Supply System," *Journal of Water Resources Planning and Management*, ASCE, Vol. 124, No. 6, November/December, pp 320-329.

GMRA (2006). <http://www.gmmra.org> (December, 2006).

Loucks, D.P., and Pallas, P. (1997). "Cost-Effective Policies for Meeting Water Demand Targets From Desalination and the Great Man-Made River Project in Libya," report submitted to Division of Policy Analysis and Operations, UNESCO, Paris, France, March, 203 pp.

Loucks, D.P. (1994). "Sustainability: Implications for Water Resources Planning and Management," *Natural Resources Forum* (special issue on sustainable development), Vol. 18, No. 4, November, pp. 263-274

CHAPTER 4

RESTORATION CASE STUDIES

Unlike the previous case studies, this chapter will focus on a critical water resources need that was essentially ignored a generation ago. Environmental restoration is now one of the most important aspects of water resources planning and management. A more ecologically focused mode of planning and management has taken root at all levels of government. The following case studies do not focus on just one aspect of planning and technology, but on the entire ecological system. The first two case studies focus on Central and South Florida. The first case study focuses on the restoration of the Kissimmee River; the second focuses on the important system that the Kissimmee River feeds—the Everglades. The case studies are linked, although the Kissimmee River case study highlights a more traditional use of technology (physical modeling) while the Everglades case study focuses on a wide variety of computer-based modeling techniques. The final case study focuses on the Louisiana Coastal Area, carefully detailing plans for restoration and discussing such key concepts as adaptive management and stakeholder involvement.

The Kissimmee River Restoration

Introduction

The Kissimmee River Basin is a large part of the headwaters of the South Florida hydrologic system and the Everglades. The Kissimmee River Basin feeds water from the Central Florida chain of lakes south to Lake Okeechobee and ultimately to the Everglades and Florida Bay. To facilitate the settlement of South Florida and eliminate catastrophic flooding in the Kissimmee River Basin, Congress authorized the United States Army Corps of Engineers (USACE) to improve the drainage of the Kissimmee River Basin. Following Congressional authorization, the USACE planned and designed the Kissimmee channelization between 1954 and 1960. Following construction, it was realized that environmental impacts as a result of the channelization were significant. This case study examines the history of the Kissimmee River and how modeling played an important role in the restoration of the River.

Background

People began moving to South Florida as early as the 1800s. What they found was a largely untamed wilderness. As the population in South Florida grew, people found themselves in conflict with the natural system. As early as the mid-1800s, Florida's government was calling for drainage of South Florida to accommodate settlement. In the early 1900s, small drainage projects began around Lake Okeechobee, which opened the area to agricultural interests south of the Lake. Settlers began moving into the area, increasing human interaction with the ecosystem.

42

The first catastrophe occurred in 1926 when a large hurricane, *The Big Blow*, moved through the southern part of Florida, killing between 325 and 800 people. This hurricane demonstrated the power of water in South Florida. In 1928 the citizens of South Florida again found themselves in the eye of a hurricane which killed nearly 2,000 people when Lake Okeechobee breached its banks, causing catastrophic flooding. These hurricanes demonstrated the need to control and manage the South Florida ecosystem to create a habitable and safe environment for settlement. In 1929 the State Legislature created the Okeechobee Flood Control District to work with the USACE on Flood Control. However, South Florida was an ever-changing system and between 1931 and 1945 the area experienced extreme drought. This period of drought was immediately followed by the heavy rainfall between 1945 and 1947. As shown in Figure 9, this rainfall caused flooding throughout South Florida and was particularly severe in the Kissimmee River Basin.

Figure 9. Flooding in the Kissimmee River Basin (SFWMD 2006b).

In 1948, Congress passed the Central and Southern Florida Flood Control Project as a result of the extreme flooding in 1947. This project paved the way for control and management of the South Florida ecosystem. Immediately following formation of the Central and Southern Florida Flood Control District, projects were proposed that would control water levels, conserve wildlife, provide flood control, and prevent saltwater intrusion. During the 1950s and the 1960s, the project was expanded to provide recreational opportunities, increase water conservation and storage, improve conveyance, and provide additional flows to Everglades National Park. The projects constructed included nearly 1,600 kilometers (1,000 miles) of canals, almost 1,600

43

kilometers (1,000 miles) of levees, more than 200 control and diversion structures, and 30 pumping stations. Of the features constructed to control levels in Lake Okeechobee, four primary canals drained water southeast towards the Lower East Coast, one canal drained waters west to the gulf, and one canal drained east to the ocean.

From the 1950s to the present, the results of the drainage of South Florida have become increasingly apparent. As a result of draining the area immediately south of Lake Okeechobee, the highly organic soils have subsided with exposure to the atmosphere. In some areas this has meant a loss of more than 1.8 meters (6 feet) of soil. The effects have been realized not only in the areas drained but also in the coastal estuaries that receive water from Lake Okeechobee. As a result of the water conveyed from Lake Okeechobee east and west through the Caloosahatchee and St. Lucie Canals, the estuaries experienced large fluctuations in salinity. North of Lake Okeechobee, the channelization of the Kissimmee River caused loss of floodplain habitat and collapse of sport fisheries.

The South Florida Water Management District (SFWMD) was created out of the Central and Southern Florida Flood Control District with the Florida Water Resources Act of 1972. As the USACE's partner in the Central and Southern Florida Project, the SFWMD was responsible for operating and maintaining many components of the Central and Southern Florida Project. The SFWMD also continues to play the key role in the management of South Florida's water resources.

To correct the problems caused by the drainage of Central and South Florida, the Central and Southern Florida Project was altered in the early 1990s. In 1992 the Water Resources Development Act authorized modifications to the Central and South Florida Project for ecosystem restoration of the Kissimmee River. This marked the advent of ecosystem restoration as a primary objective of projects in South Florida. The Comprehensive Everglades Restoration Plan was approved in 2000 under the Water Resources Development Act to "restore, protect, and preserve the water resources of central and south Florida" (CERP 2006). Included within CERP are more than 60 projects designed to improve the quantity, quality, timing of flows, and distribution of water in South Florida. The CERP is a partnership between the USACE and the SFWMD to "get the water right" in South Florida. These efforts focus on returning the South Florida ecosystem to a desirable hydrology, one beneficial to the people and the environment of South Florida.

Kissimmee River Channelization

Between 1962 and 1971 the USACE converted the Kissimmee River from a meandering 166-kilometer river (103 mile) characterized by oxbows (Figure 10) and floodplains to a straight 90-kilometer (56-mile) channel tightly controlled by six structures along its length. This method of control was selected by the USACE because it proved to be the most cost-effective project for flood damage reduction (USACE 1992).

Design and planning for channelization were carried out by the USACE after Congressional authorization was received in 1954 and construction continued until 1960. Channelization of the river occurred from 1962 to 1971. During this period the USACE dredged and cut the new channel and deposited backfill on nearly 3,250 hectares (8,000 acres) of adjacent land. This channelization severed many of the original oxbows, leaving disconnected systems that became choked by vegetation (Figure 11) and were rendered useless for foraging by fish (USACE 1992).

Because of the channelization, more than 8,100 hectares (20,000 acres) of wetland and floodplain ecosystem were eliminated, which led to reductions in the utilization of the area by certain fish and bird species. Waterfowl were especially affected and a 92% reduction in wintering bird populations was observed (SFWMD 2006a).

The Call to Restoration

Following the completion of the C-38 Canal (the Kissimmee River) in 1971, the effects of the project were quickly realized. Shortly after the channel was completed in 1971, Governor Reuben Askew convened the Conference on Water Management in South Florida. This group recommended that the "Kissimmee lakes and marshes should be restored to their historic conditions and levels to the greatest extent possible to improve the quality of water entering Lake Okeechobee" (Shen et al. 1994). In April 1978, at the request of the Committee on Public Works and Transportation (House of Representatives) and the Committee on Environment and Public Works (Senate), the USACE examined the alternatives for restoration. The USACE developed several alternatives and presented them to the Assistant Secretary of the Army for Civil Works in 1985. The USACE's report recommended no action because no alternative provided "positive net contributions" to the nation's economy (USACE 1992).

In 1983, Governor Bob Graham assembled a state council to address the problems that had resulted from the channelization of the Kissimmee River. This council included representatives from the SFWMD, the Department of Environmental Regulation, the Department of Transportation, the Department of Natural Resources, the Game and Freshwater Fish Commission, and the Department of Agriculture and Consumer Services. On this council, the SFWMD was given the responsibility of demonstrating the economic and physical feasibility of restoring the Kissimmee River and its floodplain (Shen et al. 1994). In 1985, to examine potential benefits of reconnecting lost floodplain, the SFWMD constructed three steel weirs within Pool B to re-divert water back into the original meandering channel. The result of re-inundating floodplain areas was increased catches of desirable sport fish species, increases in wading bird populations, improved wetland function, and increases in invertebrate species (USACE 1992). Because of the success of the demonstration project, the SFWMD contracted with UC Berkeley to perform mathematical and physical modeling of several restoration alternatives (Shen et al. 1994).

Figure 10. Kissimmee River, Pre-Channelization (SFWMD 2006b).

Figure 11. Choked Remnant River Reach, Post-Channelization (SFWMD 2006b).

Kissimmee River Restoration Alternative Selection

To restore the Kissimmee River, the USACE and the SFWMD evaluated four alternatives and a "no action" option. All of the alternatives provided certain system requirements as their base, including providing flood protection for the upper Kissimmee River Basin, increasing flow capacity in specific areas to rehydrate lost floodplain, and keeping the channel intact in several locations to maintain bridges.

Another similarity among all of the alternatives was the same central goal of re-inundating lost floodplain by breaching the channel and allowing water to flow through the remaining and newly constructed channel. This central goal led to the development of several alternatives, including a weir plan, a plugging plan, a Level 1 backfill plan, and a Level 2 backfill plan to achieve the desired outcome.

The weir plan recommended the use of 10 metal weirs to block the river and force flow back into adjacent river channel. These weirs would either be fixed or gated and would be set to force minimum flows through the original river channel, but in high flows they would allow overflow to prevent flooding. This plan also called for retaining the six water-control structures built as part of the channelization.

Plugging of the channel was similar to the weir plan with a recommendation for 10 plugs in the same location as the 10 weirs. Instead of a structure, each plug would be constructed from dredge material that was excavated in the initial channelization of the River. This material would be used to construct blockages in the channel that would then be reinforced and protected from erosion by rip-rap in vulnerable areas.

Level 1 backfilling of the channel proposed backfilling 10 segments of the River. This alternative maintained the structures, but in the place of a relatively small plug or weir the entire segment of channel would be filled. In this case no emergency bypass through the channel would exist. This plan also kept all of the water-control structures and would route water through these by allowing flow in the excavated channel in these areas.

The final alternative, Level 2 backfilling, called for type of filling similar to that in the Level 1 backfilling. The primary differences in the Level 2 backfilling would be the elimination of three of the water-control structures—S-65B, S-65C, and S-65D—and the construction of a new channel to connect original river sections. The extent of the backfilling would also be more significant and would include approximately 40 to 48 kilometers of canal (USACE 1992). The backfilled area would be primarily in the central reach of the canal and would completely eliminate the canal.

Kissimmee River Modeling Efforts

To model the Kissimmee River restoration, the hydrology, ecology, hydraulics, and current constraints on the system were evaluated. To examine the hydrology of the system, rainfall records were examined to determine average and extreme conditions.

Because of the complexity of the ecological system, experts recommended that the system be restored to pre-channel conditions. Unfortunately, this was infeasible because of the tight constraints on Lake Kissimmee as a result of development in the area. Following extensive review and collaboration, including a symposium, between stakeholders, it was decided that it would be necessary to re-divert flow to the original channel and allow frequent inundation of the floodplain (Shen et al. 1994).

Additional hydrology requirements were discussed in the symposium and it was decided that restoration of the pre-channelization hydrology was necessary to meet environmental restoration and ecological integrity. Of particular importance was achieving continuous flow with duration and variability similar to pre-channelization (Shen et al. 1994). Based on this goal, criteria that were developed included the following (Shen et al. 1994):

1.   Average flow velocities between 0.24 and 0.55 meters per second.
2.   Flow velocities greater than 0.24 m/s in more than 60% of cross-sections.
3.   Overbank flow along most of the floodplain when flows exceed 39.6 – 56.6 cubic meters per second.
4.   Stage-recession rates less than 0.3 meters per month.
5.   Inundation with similar frequency and period to pre-channelization hydrology.

Based on the above criteria, UC Berkeley modeled three of the four alternative restoration plans: the weir plan, a Level 1 backfill, and a Level 2 backfill.

Of particular importance in mathematical modeling is the selection of appropriate parameters. This is true when modeling a channel or overbank flow through a floodplain. In this case, selection of appropriate roughness coefficients is of particular importance. For selection of roughness coefficients, data from the field discharge test data, data from Boney Marsh, and historical data were used (Shen et al. 1994). Roughness coefficients were also examined as part of a sensitivity analysis to determine the effect of an incorrect assumption in this parameter.

For mathematical modeling of flood levels and extent in the channel, a one-dimensional network steady-flow model was applied. In general, a two-dimensional model with unsteady flow is required to measure the extent of flooding and the stage recession rate (Shen et al. 1994). However, for the Kissimmee River UC Berkeley could apply the simpler model and achieve approximate solutions that met their needs of determining flooding levels for peak flows. For calibration of the model, historical (pre-channelization) flows were used (Shen et al. 1994).

UC Berkeley's Kissimmee River Restoration modeling also included physical modeling of the river. A physical model can serve several important purposes in modeling including the appropriate duplication of a field parameter and allowing for three-dimensional unsteady-flow phenomena. The 18.3-meter-wide (60 feet) and 24.4-meter-long (80 feet) model was constructed at the Richmond Field Station in California. The model was designed so that the vertical scale was 1:40 and the horizontal scale was 1:60 (Shen et al. 1994). For the modeling effort an oxbow between S-65A and S-65B was chosen based on a reach that was sufficiently long without being overly complex (Shen et al. 1994). To construct the model, templates were made for each cross-section and were placed to the appropriate elevation with a transit. Stakes were placed approximately every 2.1 meters (7 feet) and sand was added and graded to the ground-surface elevation. To model flow outside of the channel, expected to be of high roughness, a rubberized horsehair mattress material was used to simulate approximately the same roughness as flow through the floodplain (Shen et al. 1994). Following completion of the model, calibration was performed and proved successful (Shen et al. 1994). Figure 12 shows the physical model developed for the modeling effort.

Figure 12. Physical Model for the Kissimmee River (SFWMD 2006b).

Kissimmee Modeling Results

Following development of a one-dimensional mathematical model and a three-dimensional physical model, the alternative restoration scenarios were evaluated by the UC Berkeley researchers. Based on the modeling scenarios it was determined that, in the weir plan, velocities in the oxbows would probably be sufficient to cause significant erosion that would require maintenance following flood events. Additional limitations of the weir plan would be the rapid recession of flood waters following events. Rates up to 0.30 meters per 6 hours could even occur (Shen et al. 1994).

Modeling for Level 1 backfill showed that flow velocities in the channel were still above desired levels. This would mean that maintenance would be necessary after large storm events to repair eroded areas. As in the weir plan, recession rates were too rapid because of limitations in drawdown rates at the structures in the central reach (Shen et al. 1994).

Level 2 backfilling showed water velocities similar to the desired ecological criteria. These velocities would also be unlikely to require significant repairs following floods (Shen et al. 1994).

Impacts of Modeling Efforts

Modeling at the University of California, Berkeley provided results that helped evaluate alternative restoration scenarios. Based on the results of this modeling effort, a Level 2 backfilling plan received further consideration and ultimately was selected as the preferred alternative. Following selection of this plan by the SFWMD, the USACE initiated a second feasibility study to evaluate modifications to reduce the costs of the restoration efforts (USACE 1992). Based on the modifications by the USACE and final modifications and approval by SFWMD, the costs of the Kissimmee River Restoration were estimated to be $578 million (2004 dollars) (SFWMD 2006c).

The use of physical modeling for evaluation of the restoration alternatives is an infrequently used procedure because of the cost and difficulty of such modeling. Its use in this case was compelling and valuable to determine the expected discharge values and flood levels (Shen et al. 1994). For the physical model to be effective, the selection of appropriate parameters was paramount to the quality of the outputs. To ensure the selection of appropriate values the model was calibrated. This calibration provided adequate results in comparison to historically recorded values. One important component of the modeling effort was the pairing of a physical and mathematical model. The mathematical model allows for calculation of many scenarios quickly while the physical model allows for a physical comparison to ensure that the mathematical model is providing accurate results.

Modeling efforts in this alternative selection proved important and contributed to the final selection of a restoration plan. Continued monitoring and modeling are occurring to evaluate the success of restoration efforts and the combination of operations that can provide the best holistic operating procedure within the constraints of the system. The modeling of operations known as the Kissimmee Basin Modeling and Operations Study (KBMOS) is ongoing with Phase 1 having been completed in 2005 (Earth Tech 2006).

References

Earth Tech. (2006). *Kissimmee Basin Project*, Earth Tech website, Long Beach, California, https://projects.earthtech.com/sfwmd-kissimmee/.

Shen, H. W., Guillermo III, T., and Harder, J. A. (1994). "Journal of Water Resources Planning and Management." *Kissimmee River Restoration Study*, Vol. 120, No. 3, 330-349.

South Florida Water Management District (SFWMD). (2006a). *Kissimmee River Restoration Past and Present*, SFWMD website, West Palm Beach, Florida, <http://www.sfwmd.gov/org/erd/krr/pastpres/3_krrpp.html.>

South Florida Water Management District (SFWMD). (2006b). *Kissimmee River Restoration Photo Gallery*, SFWMD website, West Palm Beach, Florida, <http://www.sfwmd.gov/org/erd/krr/photo/3_krrpg.html.>

South Florida Water Management District (SFWMD). (2006c). "Technical Publication ERA 432A." *Kissimmee River Restoration Studies: Executive Summary*, West Palm Beach, Florida.

United States Army Corps of Engineers (USACE). (1992). *Kissimmee River Restoration Study*, U.S. Gov't Print. Off., Washington D.C.

Greater Everglades Restoration Plan

Introduction

The water management system in south Florida encompasses 18,000 square miles and consists of over 1800 miles of levees and canals, about 200 major water control structures, and nearly 30 pump stations. The system, as it is today (Figure 13), can no longer effectively provide for environmental and water supply needs of the current population; therefore, it requires modification to address the needs of the predicted increase of population from about 6 million to 12 to 15 million. Beginning in 1992, the USACE, and the SFWMD initiated a "Restudy" to determine whether modifications to the existing projects and their operations are needed for improving the quality of the environment, protecting the aquifers; ensuring the integrity, capability, and conservation of urban water supplies; and maintaining flood protection (USACE and SFWMD 1999).

Unprecedented efforts are under way today to restore the Greater Everglades Ecosystem. The ongoing Kissimmee River restoration is expected to restore 27,000 acres of floodplain wetlands, reconnect 43 miles of river channel, and benefit over 300 species of fish and wildlife. Lake Okeechobee & Estuary Recovery (LOER) includes numerous projects to improve water quality inputs to the Lake from the tributaries as well as the environmental quality of the in-lake ecosystem while protecting the estuaries from damaging large discharges from Lake Okeechobee. Another major initiative, the Comprehensive Everglades Restoration Plan (CERP) (USACE and SFWMD 1999), is designed to meet water management and environmental needs over the next 50 years while meeting the needs of the urban and agricultural sectors in the region. It is expected to require more than 30 years to complete construction of facilities at an estimated cost of $10.5 billion. The State of Florida has implemented a major boost for the Everglades Restoration by expediting the pace of eight restoration projects through a program known as Acceler8.

Reliance on new technologies has been a key feature in the major projects associated with the restoration of the Greater Everglades ecosystem. Computer modeling has been a critical component in the development and implementation of the Everglades plan. This plan needed to consider agricultural, environmental, and urban water needs and, as a consequence, a multi-disciplinary approach was necessary for modeling alternatives. This case study describes the evolution of modeling practices in South Florida and how these practices have been employed for developing and implementing the Everglades Restoration plan. The next generation of tools needed for water resources planning and management in South Florida is also discussed.

Early Modeling Tools

With the advent of digital computers, SFWMD, as the primary agency for management of water resources in South Florida, in the late 1960s took its first steps in using mathematical modeling techniques (Sinha 1969). Although the early work of computer modeling focused on the Kissimmee Basin using a water budget approach, it also included simulation of groundwater aquifers using digital, analog, and hybrid computers (Khanal 1978). The multi-objective nature of water management problems in South Florida was recognized early and was analyzed through the use of operations research techniques such as Linear Programming for balancing operational costs of various water management schemes with physical and management constraints associated with several basins in South Florida (Shahane 1979).

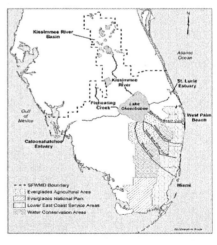

Figure 13. Left: Historic watershed that provided the basis for establishing District boundaries. Right: Major Features of the modern South Florida Water Management District.

One of the major highlights of the use of computer modeling for water resources planning in South Florida was the development and application of analog computers with input from digital simulation tools (Appel 1973; Shih and McVeigh 1978). This approach used the similarity of the unsteady groundwater flow equation and the equation describing the flow of current in an electrical circuit to build an analog computer consisting of resistors and capacitors to solve for equivalent head in a groundwater aquifer. In the electrical analog, the storage coefficient was simulated by electrical capacitance and the transmissivity by electrical conductance, with voltage and current drivers to stress the system in terms of head and flow respectively. In the model developed for a portion of the surface water/groundwater system in South Florida (Figure 14), a square grid with a resolution of 1 mile was simulated with a network of resistors and capacitors. A digital computer with a digital-analog converter was used to control the analog simulator. The model was a useful tool for drought analysis, wellfield development, seepage estimation, and salt water intrusion studies (Shahane 1979).

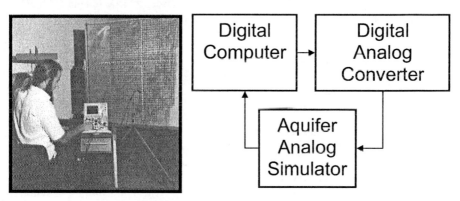

Figure 14. The electrical-circuit board of the analog model of southeastern Florida developed during the early 1970s and the primary components of the hybrid system

Regional Hydrologic Simulation Models

Because of the flat topography, high water tables, porous soils, and high transmissivity of the aquifer system, together with the extensive water control system now in place, the unique hydrology of South Florida makes the area's water management system one of the most complex in the world. This system is characterized by strong interaction between physical systems such as surface water and groundwater and among the complex management rules of various sub-regions. Regional modeling tools which capture such complex interactions are essential for planning future infrastructure changes and the operation of the entire system. A variety of hydrologic, water quality, and ecosystem models have been developed and used to analyze plans and develop operating rules for the Greater Everglades System.

Regional Routing Model. One of the early regional-scale hydrologic simulation tools was a lumped, inter-connected system model for South Florida to help water managers evaluate management policies with respect to environmental concerns, agricultural needs, urban water use, and flood protection requirements (Trimble 1986). The model predicts areal averaged water levels and major structure discharges for Lake Okeechobee and Water Conservation Areas under different management options and historical hydrologic conditions. A simple mass balance approach is used to simulate water levels and discharges associated with each system component, conceptualized as a storage area. The model also requires pre-processing of water demands of Lake Okeechobee and Lower East Coast urbanized areas, developed using other, more sophisticated models. Constraints on water movement between system components are incorporated into the model. The model has been used extensively for analyzing planning and operations (e.g., the operation of Lake Okeechobee).

Everglades Screening Model (ESM). Derived from the basic concepts used in the Regional Routing Model, the Everglades Screening Model is a mass balance model that simulates the major hydrologic features and demands of the South Florida water resources system. The model, used extensively during the screening phase of the Everglades Plan, was developed using the STELLA Object-Oriented modeling environment and runs on a weekly time step. The model simulates most operating rules associated with major impoundments and water control structures. The strength of the ESM is its ability to evaluate water management components and/or alternatives in a relatively short time.

South Florida Water Management Model (SFWMM). The South Florida Water Management Model is an integrated surface water–groundwater model that simulates the hydrology and existing or proposed water management plans in the South Florida region using climatic data for the 1965-2000 period (SFWMM 1999). The model simulates the major components of the hydrologic cycle in South Florida, including rainfall, evapotranspiration, overland flow, groundwater flow, canal flow, and seepage across levees. The model also simulates operation of the water management system components, including major well-fields in the urbanized east coast, impoundments, canals, pump stations, and other water-control structures. Two-dimensional regional hydrologic processes such as overland flow and groundwater flow are simulated in the model at a daily time step using a mesh of 2-mile-x-2-mile grid cells.

The ability to simulate key water-shortage policies affecting urban, agricultural, and environmental water uses allows the users of SFWMM to investigate trade-offs among different water-management objectives. The model produces extensive output throughout the system, which can be summarized into numerous performance measures and indicators for evaluating water-management plans. The SFWMM is a premier hydrologic simulation model used for system-wide evaluation of Everglades Restoration plans.

Natural System Model (NSM). The Natural System Model (SFWMD 1998) attempts to simulate the hydrologic response of the pre-drainage Everglades using the same climatic inputs, daily time step, calibrated model parameters and algorithms as the SFWMM. The NSM differs from SFWMM in that it does not simulate the influences of any man-made features and uses estimates of pre-subsidence topography and historical vegetation cover. The NSM does not simulate the particular hydrologic conditions that existed before human influence in South Florida, but rather its hydrologic response due to recent climatic response. This allows the modelers to compare the performance of the managed system with that under natural conditions. The NSM has played a major role as a tool for setting environmental restoration targets for the Comprehensive Everglades Restoration Plan.

Everglades Landscape Model (ELM): The Everglades Landscape Model (Fitz et al. 2004) belongs to a new and innovative class of models recently developed to combine hydrologic, water quality, and ecological processes in a single model to predict spatial and temporal patterns of landscape changes (Figure 15). The model explicitly simulates interactions and feedbacks between water, nutrients, soils, and wetland plant dynamics using a mesh consisting of approximately 10,000 square cells of 1 km^2 each in the greater Everglades regional application. The purpose of ELM was to (a) integrate hydrology, biology, and nutrient cycling in spatially explicit, dynamic simulations; (b) provide a framework for collaborative field research and other modeling efforts; and (c) understand and predict long-term relative responses of the landscape to different management scenarios.

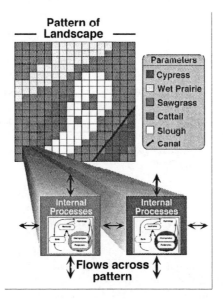

Figure 15. The ELM ecosystem dynamics applied across a heterogeneous grid of habitat types.

55

Each habitat type within the patterned landscape may be parameterized differently, affecting the internal process dynamics within grid cells. In turn, the results of the internal processing affect the flows of water and nutrients across the landscape pattern. Habitat succession occurs as cumulative conditions warrant.

The model requires daily data for all managed water control structure flows, either measured or predicted by SFWMM, and predicts the spatial and temporal variation of macrophyte and periphyton biomass and community types, soil elevation, phosphorus concentrations, and water levels within the model domain. The general model is scalable and it has been applied to smaller sub-regions within the Everglades.

Across Trophic Level System Simulation (ATLSS) Model: Belonging to the same class of new generation of models as ELM, ATLSS is designed to be an integrated system of simulation models representing the biotic communities of the Everglades/Big Cypress region and the abiotic factors that affect them (DeAngelis et al.1998). The models are spatially explicit and have a resolution of 500 m x 500 m or finer. The abiotic components simulated by the models include methods to rescale hydrologic outputs from other models and a fire process model. The ATLSS models use the output of the SFWMM or any other hydrologic model. The biotic modeling components integrate three approaches: (a) Spatially-Explicit Species Index models, which compute indices for breeding or foraging success of various animal species; (b) structured population models for several important functional groups of fish and macroinvertebrates; and (c) individual-based models for large consumers (wood storks, great blue herons, white ibis, American alligators, white-tailed deer, and Florida panther). The overall goal of the ATLSS is to aid in understanding how the biotic communities of South Florida are affected by the hydrologic regime and other abiotic factors and to provide a predictive tool for evaluating management alternatives.

River of Grass Evaluation Methodology (ROGEM): The ROGEM is a collection of nine community-level equations which were developed to predict relative habitat quality responses to Everglades restoration alternatives. Equation outputs represent quality of fish and wildlife habitat on a 0 to 1 scale. Most equation variables are hydrologic and the information for executing ROGEM is obtained from the output of the SFWMM for a particular alternative scenario.

Conceptual Models: The conceptual models are simple, non-quantitative tools for organizing and communicating existing, empirically, and intuitively derived understandings of key physical and biological relations of ecosystems (Ogden et al. 1997). These models link the following major elements: (a) societal drives or sources (e.g. urban and agricultural expansion), (b) major stressors acting on the system (e.g. altered hydropattern), (c) major ecological effects from the stressors, (d) ecological attributes (indicators), and (e) recommended measures for each attribute which collectively reveal the health of the ecosystem. The conceptual models have been developed for several major landscapes of the Everglades and they represent the collective understanding of the landscapes and the interpretations of past studies and hypotheses discussed through numerous workshops.

Role of Models in Everglades Plan

Development of the Comprehensive Everglades Restoration Plan was undertaken by a large, inter-agency, multi-disciplinary team lead by the co-sponsoring agencies, the USACE, and the SFWMD (Tarboton et al. 1999). The process was an exercise in Shared Vision Modeling, which included a complex, iterative process involving the models described above at several stages (Figure 16). An initial screening process, consisting of numerous runs of ESM, a Cost-Effectiveness Analysis, and the findings from the previous regional water supply planning efforts, was designed and executed to begin the process of alternative development. Two major sub-teams, one for alternative development and the other for alternative evaluation, were formed from the interdisciplinary study team which consisted of members from multiple agencies and stakeholder groups. Before modeling alternatives, a baseline condition representing 1995 land use and demands and a future without-project condition representing projected 2050 land use and demands were defined.

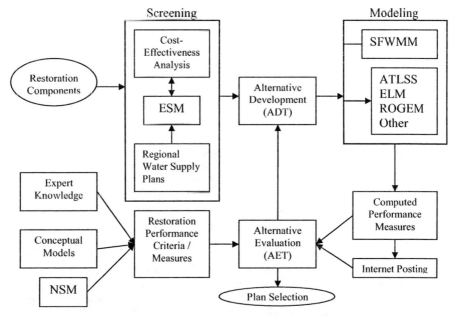

Figure 16. Role of models in the planning process used in the CERP development process.

The alternative design team (ADT) designed and modeled features of each alternative in the SFWMM. Performance Measures, computed directly from SFWMM output, were published on the internet to facilitate the review of each alternative by the alternative evaluation team (AET) and to communicate alternative results to interested agencies, stake-holders and the public. Output from the SFWMM provided input to the water quality, ecological, and species-specific models. The AET incorporated comments received from different agencies and the public on each alternative together with their own evaluation to make recommendations to the ADT for refinements to improve subsequent alternatives. An independent cost analysis was undertaken for each alternative and the preferred alternative was selected following a comprehensive evaluation of all alternatives.

Conceptual Models, expert knowledge of the scientists of the AET, and the Natural System Model (NSM) played major roles in the definition of Performance Criteria and Measures. Internet posting was the primary means of communicating modeling results to agency partners and stakeholders. There were over 900 tables and graphics for each set of alternative comparisons made, which demonstrated that the internet was a very efficient way to share this information with a large audience.

The Comprehensive Everglades Restoration Plan, which consists of more than 60 components designed to achieve a balance between ecosystem restoration and urban and agricultural water supply, has been described as the world's largest ecosystem restoration effort.

Operational Planning

Advances in the ability to predict future climatic regimes have allowed this science to become a plausible mechanism for achieving more efficient regional water management in South Florida (Trimble et al. 2006). The best example of the application of climate outlook for water management in South Florida is the operational schedule of Lake Okeechobee. Traditional rule curves (or regulation schedules) for operation of a typical reservoir are "static," with decisions made only when the reservoir level crosses a predetermined curve. In 1998, a climate-based regulation schedule for Lake Okeechobee was developed by incorporating not only the seasonal and multi-seasonal climate outlooks but also the near-term (two-week) forecasts of tributary inflows into the lake. The new schedule, known as the *Water Supply and Environment* (WSE), attempts to balance the multiple objectives of managing Lake Okeechobee. The overall operational strategy in developing the WSE schedule was to improve the performance with respect to five water management objectives associated with (a) maintaining flood protection; (b) minimizing urban and agricultural water supply shortages; (c) minimizing damaging estuary discharges (c) improving the Everglades hydroperiod; and (d) improving the in-lake, littoral zone hydroperiod. Because of the competing nature of some of the objectives, a multi-objective trade-off analysis was conducted by using SFWMM.

The final regulation schedule for Lake Okeechobee consisted of two main components: (a) a set of regulation schedule lines that define different operational zones and (b) decision trees that support the process for making discharges based on forecasts of inflows and climate outlook (Figure 17). The climate-based operational guidelines, as incorporated into the WSE regulation schedule, have emerged as a highly desirable approach for Lake Okeechobee water management.

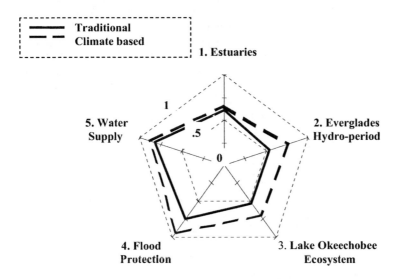

Figure 17. Normalized Multi-Objective Trade -Off Analysis:
Traditional versus climate-based operations (0 = worst, 1 = best).

Implementation of climate-based schedule: The WSE schedule was adopted in July 2000. The weekly implementation of this operational schedule requires the following information: (a) current water level, (b) tributary hydrologic condition, (c) seasonal (six months in duration), and (d) multi-seasonal (seven to 12 months in duration) climate outlook in terms of expected Lake Okeechobee Net Inflow. The USACE (2000) water control manual describes the methods which are used to estimate the required parameters to follow the decision trees. Every week on Tuesday, after the hydrologic and environmental data are collected and analyzed on Monday, an interdisciplinary team of scientists and engineers from state and federal agencies meet to review the status of the regional system and the implementation of the WSE schedule.

Position Analysis: Seasonal and multi-seasonal operational planning of major water resources systems requires a careful evaluation of likely future scenarios of water and environmental conditions that influence management objectives. The SFWMD uses Position Analysis using SFWMM as a form of risk analysis that can forecast uncertainties associated with a specific operating plan for a basin over a period of many months conditioned on the current state (e.g., reservoir storages) of the system (Hirsh 1978; Cadavid et al. 1999). The District relies on generating a large number of possible traces with durations of one or more seasons for the hydrologic variable of interest (e.g., reservoir stages or flows), using the same initial conditions and broad ranges of meteorological conditions that may occur in the future but cannot be forecast accurately.

During the past decade, position analysis concepts developed by the District have been used with increasing effectiveness to assess risks associated with seasonal and multi-seasonal operations of the water management system and communicate the projected outlook to decision makers, agency partners, stakeholders, and the public. While the SFWMD has used position analysis chiefly to project the expected stage of Lake Okeechobee, it has also used the technique for other impoundments, including the WCAs. Monthly position analysis has become an important tool for making operational decisions that may have implications for multiple seasons.

Next Generation Modeling

Efforts are underway to replace the legacy regional modeling tool, SFWMM. Recognizing the need for a next-generation regional modeling tool, SFWMD has embarked on a major development effort to develop the next generation regional modeling tools, Regional Simulation Model (RSM). The RSM has two principal components (Figure 18), the Hydrologic Simulation Engine (HSE) and the Management Simulation Engine (MSE).

The HSE component solves the governing equations of water flow through the land-phase of the hydrologic cycle and the man-made features within the model domain using a finite-volume approach implemented on a triangular network of mesh elements. The MSE component provides a variety of system management capabilities by implementing operating rules for both water supply and flood control, regulation schedules (e.g. rules for a reservoir), operation of water control structures, and their coordination within the model domain. The development of RSM has relied mainly on the following building blocks: new computational methods (Lal 1998; Lal 2000), object-oriented (OO) code design and implementation, new and efficient numerical solvers for large matrices (SFWMD 2005), and the experience of regional modeling in South Florida gained from the SFWMM.

In the RSM, the Hydrologic Simulation Engine (HSE) provides hydrological and hydraulic state information, Σ, while operational policies dictate managerial constraints and objectives, Λ. In the MSE this state and process information can be functionally transformed or filtered by Assessors (A). The MSE then produces water management control signals (χ,μ) which are applied to the hydraulic control structures to satisfy the desired constraints and objectives. Figure 19 illustrates this overall cyclic flow of state and management information in the RSM.

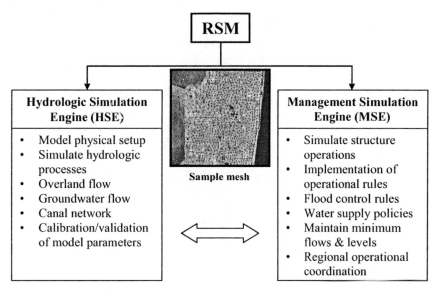

Figure 18. Principal components of the next generation regional model, RSM, and sample mesh for an urbanized area in the Lower East Coast of Florida.

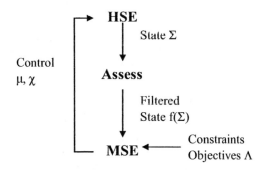

Figure 19. RSM state and management information flow.

Hydrologic Simulation Engine (HSE)

The RSM approach differs from the traditional approach of model development in many ways.. The design of the model captures the essential theoretical framework necessary to simulate unique features of South Florida hydrology while using an object-oriented framework consisting of "water bodies" and "water movers" as described below.

The governing equations (mass balance and conservation of momentum) for simulating hydrology of the physical system are derived from the Reynolds transport theorem (Chow et al. 1988). In arriving at the solution to the governing equations (mass and momentum), simplifications were made for RSM by assuming inertia terms to be negligible for the intended applications, i.e., continuous simulations for planning purposes (Lal 2001). The resulting diffusive wave formulation has allowed the RSM to combine the surface water and groundwater flow equations which can be solved using a single matrix.

The object-oriented design using C++ programming language has evolved into two basic abstractions of the fundamental elements included in the governing equations: (a) "waterbodies" to represent storage in mesh cells, canal segments, and lakes etc.; and (b) "watermovers" to represent water movement between waterbodies. The details of the numerical solution of the governing equations and the various forms of waterbodies and watermovers can be found in (Lal 1998; SFWMD 2005). The sources and sinks of the hydrologic system are computed using the concept of Hydrologic Process Modules (HPM), which can simulate the effects of "vertical" processes such as rainfall, evapotranspiration, recharge, irrigation, and others using highly detailed information about the landscape (SFWMD 2005).

Management Simulation Engine (MSE)

The MSE consists of a multi-level hierarchical control scheme, which naturally encompassed the local control of a particular water control structure, as well as the coordinated sub-regional and regional control of multiple structures. The RSM architecture emphasizes the decoupling of hydrological state information from the management information processing applied to the states. Given a well-defined interface between the HSE and MSE, this approach enables multiple information processing algorithms to execute in parallel, with higher levels of the hierarchical management able to synthesize the individual results which are best suited to the managerial objectives.

In the MSE, this data access is achieved by storing hydrological and managerial information relevant to a water control unit (WCU) in a data storage object defined in a MSE Network. The MSE network is an abstraction of the stream flow network and control structures suited to the needs of water resource routing and decisions. MSE networks maintain assessed and filtered state information, parameter storage relevant to WCU, or hydraulic structure managerial constraints and variables, and serve as an integrated data source for any MSE algorithm seeking current state information. They also provide a mathematical representation of a constrained interconnected flow network, which facilitates the efficient graph theory solution of network connectivity and flow algorithms.

The primary object in the MSE network is the Water Control Unit (WCU). A WCU maps a collection of HSE stream segments that are operationally managed as a discrete entity to a single WCU in the MSE network. WCUs are typically bounded by hydraulic control structures, which are represented as nodes in the MSE network. Each WCU includes associative references to all inlet and outlet hydraulic flow nodes.

References

Appel, C.A. (1973). "Electrical-Analog Model Study of a Hydrologic System in Southeast Florida" U.S. Geological Survey, Open File Report.

Cadavid L.G., Van Zee, R., White, C., Trimble, P., and Obeysekera, J. (1999). "Operational Hydrology in South Florida Using Climate Forecasts" American Geophysical Union, 19th Annual Hydrology Days. Colorado State University, August 16-20, 1999. ed. H.J. Morel Seytoux. Hydrology Days Pub.

Chow, V., Maidment, D. and Mays, L. (1988). *Applied Hydrology*. New York, NY: McGraw-Hill Book Company.

Fleming, D. M., DeAngelis, D. L., Gross, L. J., Ulanowicz, R. E., Wolff, W. F., Loftus, W. F., and Huston, M. A. (1994). "ATLSS: Across-Trophic-Level System Simulation for the Freshwater Wetlands of the Everglades and Big Cypress Swamp". National Biological Service Technical Report.

Hirsch, R. M. (1978). "Risk Analysis for a Water-Supply System – Occoquan Reservoir, Fairfax and Prince William counties, Virginia.", *HydroL Sci B* 23(4), 476-505.

Khanal, N. (1978). "Application of Computer Techniques for Long Range Regional Groundwater Resources Management," paper presented at the 10th International Higher Hydrological Course, Moscow State University, June 10-August 10.

Lal, A. M. W. (2000). "Numerical errors in groundwater and overland flow models". *Water Resources Research* 36 (5), 1237–1247.

Lal, A. M. W. (1998). "Weighted implicit finite-volume model for overland flow". *Journal of Hydraulic Engineering*, ASCE 124 (9), 941–950.

Ogden, J.S., Davis, S., Rudnick, D., and Gilick, L. (1997). Natural Systems Team report to the Southern Everglades Restoration Alliance, Final Draft, South Florida Water Management District, West Palm Beach, Florida.

SFWMD. (1998). "Natural System Model Version 4.5 Documentation," Hydrologic Systems Modeling Division, South Florida Water Management District, West Palm Beach, Florida.

SFWMD (2005). "Theory Manual, Regional Simulation Model". South Florida Water Management District, Office of Modeling, May 16, 2005.

Shahane, A. (1979). "Mathematical Models in Water Resources Planning," A Memorandum Report. South Florida Water Management District, West Palm Beach, FL.

Shih G., and McVeigh, F., (1978). "A Hybrid Computer System For Groundwater Modeling," in Verification Mathematical and Physical Models in Hydraulic Engineering," Proceedings of the 26th Annual Hydraulics Division Specialty Conference, University of Maryland.

Sinha, L. K. (1969). "An Operational Watershed Model: Step 1-B; Regulation of Water Levels in the Kissimmee River Basin," paper presented at the Fifth Annual Water Resources Conference, San Antonio, Texas, October 27-30.

Tarboton K.C., Neidrauer, C., Santee, E., and Needle, J. (1999) "Regional Hydrologic Modeling for Planning the Management of South Florida's Water Resources through 2050," paper presented at the Annual International Meeting of the ASAE, Toronto, Ontario, Canada.

Trimble, P.J., Obeysekera, T.B., Cadavid, L.G., and Santee, E.R. (2006). "Applications of Climate Outlooks for Water Management in South Florida". In *Climate Variations, Climate Change, and Water Resources Engineering*. Edited by Jurgen D. Garbrecht and Thomas C. Piechota, ASCE/EWRI. Reston, VA.

Trimble, P. (1986). "South Florida Regional Routing Model," Technical Publication 86-3," South Florida Water Management District, West Palm Beach, Florida.

USACE and SFWMD. (1999). Central and Southern Florida project, Comprehensive Review Study. Final Integrated Feasibility Report and Programmatic Environmental Impact Statement, Vol I. USACE Jacksonville FL and SFWMD, West Palm Beach FL. 586 pp.

U. S. Army Corps of Engineers. (2000). "Water Control Plan for Lake Okeechobee and the Everglades Agricultural Area," Jacksonville District, Florida.

Louisiana Coastal Area Study

Introduction

Louisiana's coastal wetlands have lost between 1,500 and 1,900 square miles in the last century. From 1930 to 1990, the coastal zone of Louisiana lost an estimated 3,950 square kilometers, or 1,526 square miles, of wetlands (Boesch et al. 1994). The Louisiana coastal plain contains one of the largest expanses of coastal wetlands in the contiguous United States, making up about 40 percent of the Nation's coastal marshes—but it has accounted for 90 percent of the total coastal marsh loss in the Nation. This wetland loss has had major adverse effects on the region's ecosystem and the wetlands are estimated to lose an additional 500 square miles over the next 50 years. This historical and continued loss has affected and will continue to significantly affect the ecology, society, and economy of the region and the Nation. As this natural ecosystem continues to decline, the result will be decreases in various natural functions and values associated with wetlands, including diminished biological productivity and increased risk to critical habitat of Federally-listed threatened and endangered species. The ability of the coastal wetlands to buffer tropical storm and hurricane storm surges will diminish, increasing the risk of significant damage to oil, gas, transportation, water supply, and other private and public infrastructure and agriculture lands and urban areas.

Numerous small-scale restoration projects constructed over the previous 20 to 30 years provided primarily localized remedies. However, this "piecemeal approach" of individual restoration projects constructed with little or no coordination nor evaluation of their role in a "coastal consistency" framework did little to solve the overall problem of the massive loss of wetlands. Given the magnitude of Louisiana's coastal land losses and ecosystem degradation, it became apparent that a systematic approach involving larger projects to restore natural geomorphic structures and processes, working in concert with smaller projects, would be required to effectively deal with the full extent of the degradation and ensure a sustainable coastal ecosystem. In 1998, state and Federal agencies, local governments, academia, numerous NGOs, and private citizens reached consensus on "Coast 2050 – Toward a Sustainable Coastal Louisiana" (LA CWCRTF 1998), a conceptual plan for restoring the Louisiana coast. The Coast 2050 Plan was a direct outgrowth of lessons learned from implementation of past restoration projects and reflected a growing recognition that a more comprehensive "systemic" approach was needed.

The Louisiana Department of Natural Resources (La DNR) and the New Orleans District US Army Corps of Engineers (USACE) initiated the Louisiana Coastal Area (LCA) study in early 2002. The primary objective of the study was to produce a comprehensive program for restoring and maintaining an ecologically sustainable ecosystem for the Louisiana coastal zone; a plan to which Louisiana and the Federal government would be willing to dedicate significant resources to achieve. Numerous Federal, state, and local agencies; various non-government organizations (NGOs); and academic institutions participated significantly in formulating the plan for the LCA study and in recommending plan development. In November 2004, the Louisiana Coastal Area (LCA)–Ecosystem Restoration Study Final Report (USACE 2004) was completed. The recommended plan called for action on several levels. The LCA report presented a strategy for addressing the long-term needs of coastal Louisiana and identified short-term needs and projects that could be implemented to slow the deterioration of the coastal wetlands while the more robust long-term features could be properly developed and implemented.

The role of technology in the recommended plan was evident in the initial development of a new modeling tool to evaluate the cumulative impacts of proposed measures on the ecosystem. In addition, the LCA study team consisted not only of

Louisiana DNR and USACE biologists and engineers, but also biologists, engineers and other specialists from numerous other agencies such as the Louisiana Department of Wildlife and Fisheries, the U.S. Geological Survey, the U.S. Fish & Wildlife Service, the U.S. Environmental Protection Agency, and various academic institutions. Incorporating the technical expertise of these agencies throughout the study, rather than near the end (more commonly done as part of the review and comment process), helped ensure that the most up-to-date technology was used to develop the recommended plan. However, acknowledging that not all the necessary knowledge and technology currently exists, the recommended plan also called for a science and technology program through the LCA project to identify, fund, and develop needed technology and knowledge for direct application to the LCA project. Also included in the plan were continued monitoring and adaptive management over the life of the project to enable future adjustments to the overall plan, as better knowledge and technology became available.

LCA Study Plan and Recommendations

In response to the continuing loss of wetlands in the Louisiana coastal zone and the concern about eventual ecosystem collapse, the State of Louisiana (through the Louisiana DNR) and the USACE initiated, in early 2002, the Louisiana Coastal Area (LCA) study. The primary purpose of the LCA study was to analyze the problems, their causes, and possible alternatives to reduce or eliminate future deterioration of the wetlands and to develop a plan to begin to rebuild the wetlands. Building upon past efforts such as Coast 2050, the intent was to produce a report, nearing a feasibility-level study, which could be processed through the Administration and presented to Congress as a basis for authorization and funding of a long-term plan for coastal restoration. The study report (USACE 2004) was completed in November 2004 and was included in both the House and Senate versions of the proposed 2006 Water Resources Development Act (WRDA). While WRDA 2006 was not completed, it is anticipated that the LCA project will be included in a subsequent WRDA. Passage of the WRDA bill will authorize the LCA plan and, with appropriations of funding, allow the LCA plan to begin implementation.

The LCA-recommended plan set priorities for near-term projects, expanding knowledge and capabilities through a 10-year Science and Technology (S&T) Program with demonstration projects and expanding the authority and funding to beneficially use more of the dredged materials taken from navigation channels. In addition, the plan provided for large-scale, long-term restoration studies and measures for which current levels of analysis and design were not adequate for deciding whether to proceed with implementation. The LCA plan, with an estimated cost of just under $2 billion, is briefly described in the following paragraphs. Both LaDNR and the USACE stress that this is just the first step in a longer process, and when additional studies and designs are completed it is anticipated that Congress will be asked to authorize and fund more major projects. The recommended plan features of the LCA Report (USACE 2004), broken down into several major categories, are listed below with their estimated 2004 costs:

Initial Near-Term Critical Restoration Features	$ 864 million
Additional Near-Term Critical Restoration Features	$ 762 million
Science and Technology Program	$ 100 million
Science and Technology Demonstration Projects	$ 100 million
Beneficial Use of Dredged Materials	$ 100 million
Modification of Existing Structures Studies	$ 10 million
Large-Scale, Long-Term Restoration Studies	$ 60 million
Total	**$1,996 million**

The initial near-term critical restoration features consist of five individual projects designed to meet critical ecological needs of the Louisiana coastal area in critical locations. Delaying action at these locations would result in continued losses in these areas and would thus require greater restoration costs when these areas were eventually addressed. Three of the projects divert freshwater from the Mississippi River into marsh areas to reduce salinity intrusion and help build marshes. A fourth project would restore a critical reach of barrier islands. The fifth project, consisting of shore protection, would temporarily address continued erosion of the Mississippi River Gulf Outlet (MRGO) channel, a deep-draft navigation channel, constructed in the 1960s, linking the Port of New Orleans with the Gulf of Mexico. Continued erosion of this channel's banks would increase salinity intrusion into this area with further losses of marshes. Normally, Congress would require the completed decision documents before authorizing the individual projects. However, a significant amount of engineering and design as well as environmental analysis has been conducted for these features from earlier efforts. The LCA plan recommends that Congress authorize these features, subject to completion of a decision document that would then be approved by the USACE Chief of Engineers. Significant time can be saved by such a programmatic authorization and construction could begin within the first five years after authorization.

The additional near-term critical restoration features consist of an additional 10 individual projects for which initial analyses have begun but have not proceeded to the point where a final decision should be made. These features include various measures such as additional river diversions of freshwater and sediment, marsh creation/restoration, and barrier island restoration designed to improve water and sediment management in the marshes. When analyses and designs of these features are completed, decision documents would then be passed to Congress for authorization and appropriation in future WRDA's.

The LCA's S&T Program and Demonstration Projects provide a mechanism to improve the science and tools necessary to adequately plan and design a plan for a sustainable ecosystem. While the knowledge and technology base of coastal ecology is substantial, scientists, engineers, and ecologists do not know everything necessary to completely design and plan ecosystem restoration. The S&T Program will have a Science Director who will bring together the appropriate academic and research elements necessary to resolve scientific uncertainties concerning restoration causes and effects and to develop the science and modeling tools necessary to reduce uncertainty about ecosystem interrelationships and project the expected benefits and impacts of proposed features. Demonstration projects will be designed to resolve uncertainties or demonstrate project effectiveness on small scales before application to larger, more costly systems.

The Beneficial Use of Dredged Material feature would provide an additional $100 million over 10 years to increase the amount of dredged material used for creating marsh. The USACE's New Orleans District currently only beneficially places for marsh restoration about 20 to 25 percent of the approximately 70 million cubic yards of material per year dredged to maintain the authorized navigation channels in southern Louisiana. Under the existing beneficial use program policies and due to limited funding for maintenance dredging, the District cannot significantly increase its dredging costs to beneficially dispose of the dredged material. Beneficial use has thus normally been limited to areas near the maintained waterways. Interior marshes or open-water areas several miles or more away from these waterways cannot normally be reached with the disposal techniques without significantly increasing costs. This program would provide additional funds for the extra costs of moving the dredged material greater distances and give greater flexibility in restoring or preserving interior marshes. Use of this program would also significantly reduce the amount of dredged material disposed of offshore with little or no environmental benefits.

The LCA plan would also include studies to identify structures that could be changed structurally or operationally to provide or improve the structures' abilities to contribute to the ecosystem, in most cases with little or no changes to the original objectives of the structures. Such changes could be used to reduce salinity intrusion or divert additional freshwater or sediments to help restore or protect threatened areas.

The LCA plan also includes large-scale, long-term studies that not only have the potential to make macro-scale changes in the ecosystem but also macro-scale changes to the existing uses of the system. An example would consist of creating a new tributary of the Mississippi River to form a new "delta" either east or west of the existing river. Such a feature could potentially divert up to one-third or more of the average flow and sediment of the Mississippi River. Such large-scale studies would have to demonstrate that the proposals were technically feasible and had major benefits to the coastal ecosystem while determining and addressing any adverse impacts such proposals would have to the users and stakeholders of the existing ecosystems, streams, and receiving marshes. Unintended adverse impacts to the coastal and riverine ecosystems from such proposals would also be addressed by these studies.

Science and Technology Program

Technology played a significant and prominent role in the LCA study. At the beginning of the LCA study, the study team recognized that, while the current science and technology knowledge and abilities relative to coastal ecosystems are substantial, there was still a need for further advancements to reduce the scientific uncertainties and expand the engineering technology for coastal restoration. To address these needs, the recommended LCA plan included a 10-year S&T Program funded for up to $100 million. A major component of the S&T Program would include demonstration projects to deepen knowledge and improve the technology for coastal restoration. The LCA S&T Program would provide a strategy, organizational structure, and process to facilitate integration of science and technology into the decision-making processes of the LCA Project Execution Teams. Implementation of the S&T Program would ensure that the best available science and technology available were used in the planning, design, construction, and operation of LCA Plan features.

Uncertainties may be related to data availability, science, modeling, and other analytical tools; socio-economic impacts; implementation; technical methodology; resource constraints; cost; or effectiveness of restoration features. Uncertainties may also be related to development and refinement of forecasting tools. Major roles of the S&T Program will be to identify and prioritize critical areas of uncertainty, to formulate the most appropriate means of resolving uncertainties, and to ensure focused data collection aimed at resolving these areas of uncertainty. Results would be used to make recommendations to the LCA program regarding program and project refinements in light of the reduced uncertainty. Critical areas of uncertainty identified by the study team, academics, or agency personnel would be proposed to the S&T Office Director. However, the S&T Office would not be constrained to targeting only these needs, but rather would be open to facilitating the pursuit of new technology, experimentation, and innovative ideas when suitable for the advancement of the LCA program. Areas of uncertainty would be prioritized based on how much resolving the uncertainty would advance the LCA Program.

The S&T program and its Director would work with the LCA program management and study team to review and assess goals and objectives of the LCA program and to identify S&T needs to help the LCA Plan meet those goals and objectives. The S&T Program would manage and coordinate science projects for data acquisition and monitoring, data management, modeling, and research to meet identified scientific needs of the LCA Plan. The program would establish and maintain independent science and technology advisory and review boards and conduct scientific evaluations, assessments, and peer reviews to ensure that the science implemented, conducted, or produced by the S&T Program meets an acceptable standard of quality, credibility, and integrity. In addition, the S&T program would coordinate with other research efforts, such as the Louisiana Governor's Applied Coastal Research and Development Program, and other state and Federal R&D entities. The program would also incorporate lessons learned and experiences (pros and cons) of other large-scale ecosystem restoration science and engineering programs such as the Everglades, Chesapeake Bay, and Calfed. The program would establish performance measures for restoration projects and monitor and evaluate the performance of program elements. The S&T program would also prepare scientific documents including a periodic Science and Technology Report and conduct technical workshops and conferences.

Through the S&T program, an improved scientific understanding of coastal restoration issues would be gained and be infused into planned or future restoration planning, projects, and processes conducted by the LCA project study team.

Demonstration projects represent one of several strategies that the S&T program would employ to reduce uncertainties. Demonstration projects may be necessary to address uncertainties not yet known and discovered in the course of individual project implementation or during studies of large-scale and long-term restoration concepts. The S&T Director would prepare documents that would identify major scientific or technological uncertainties to be resolved and a monitoring and assessment plan to ensure that the demonstration project would provide results that contribute to the overall LCA program effectiveness. After design, construction, monitoring, and assessment of individual demonstration projects, the lessons learned would be applied to improve the planning, design, and implementation of other Louisiana coastal zone restoration projects. Under the LCA program, these demonstration projects would be funded up to $100 million over 10 years, with no single demonstration projects exceeding $25 million.

CLEAR Modeling

For the LCA study, modeling tools were developed to assess the impacts—both beneficial and adverse—the various proposed restoration measures would have on the Louisiana coastal ecosystem. The knowledge of how coastal ecosystems function has grown dramatically over the past 50 years. However, it would be inaccurate to state that we know enough about how the various components of the ecosystem interact with each other and react to various natural and man-made changes. Therefore, it would be difficult to say with certainty how an ecosystem the size of the LCA study area will respond to numerous, combined, or overlapping restoration measures. Restoration projects to date have generally focused on small areas or localized ecosystems much smaller than the LCA study area.

While many of the results would be expected to carry over to a larger scale, the overlapping or cumulative impacts of many restoration measures could produce many unintended impacts. As such, the LCA study team considered it critical to develop a new modeling approach and apply it to assess the overall ecosystem response to proposed measures. A large number of academic scientists and ecologists (from Louisiana State University, University of Louisiana at Lafayette, University of New Orleans, and others), working with the other resource agencies, developed an LCA Ecosystem Model to evaluate and assess multiple combinations of restoration strategies and measures for the study. The model became known as the CLEAR model (Coastal Louisiana Ecological Assessment and Restoration).

Developing and evaluating coastal restoration features of the LCA to achieve this goal required linking the changes in environmental drivers (processes such as riverine input) to specific restoration endpoints (hydrodynamic, ecological, and water quality) using a variety of modeling approaches. The linkage of numerous proposed restoration measures and the projected results of these measures were provided by the development of the CLEAR Model. The modeling system consists of five major steps in the evaluation process. In Step One the frameworks that approximate the degree of change in environmental settings to achieve planning scales (reduce, maintain, increase, etc.) were developed. In Step Two the frameworks were provided to an ecosystem modeling team (consisting of agency and academic experts) for estimates of change in five modules: (1) hydrodynamics, (2) land building, (3) habitat switching, (4) habitat use, and (5) water quality. Each module required knowledge of existing conditions and the ability to predict changes in the landscape based on assumptions of how the ecosystems respond to coastal processes. In Step Three each module produced a set of endpoints specific to the environmental conditions of the particular coastal measures. Many of these endpoints became the input to other modules. Step Four used the endpoints of these five modules in a series of ecosystem benefit calculations to determine specific types of ecosystem response. Finally, in Step Five the original restoration frameworks were evaluated using a collection of the benefits and compared to the original restoration objectives.

The CLEAR Model was used to evaluate the cumulative impacts of the proposed restoration comprehensive plans on individual subprovince areas. Some of these comprehensive plans were used in the final array of measures for the LCA recommended plan. Many of the possible combinations predicted land creation (or at least reduction in loss rates), but often with undesired results in many of the subprovinces, such as over-freshening of the estuaries with reductions in fisheries resources. However, overall the model allowed the study team to evaluate, at least on a preliminary scale, the numerous combinations of restoration measures and their predicted impacts on the LCA ecosystem.

The CLEAR Model represents a significant advancement in the ability to evaluate coastal ecosystems. However, the model development and resolution obtained during the LCA study allowed only macro-scale estimates of how proposed comprehensive restoration plans would impact the coastal processes and provide for a sustainable coastal landscape. Future models will be developed during subsequent LCA studies to enable the evaluation of proposed measures on areal and ecosystem scales at a much finer resolution. This will allow analysis and evaluation and help reduce the scientific uncertainty of the impacts of such measures on the ecosystem linkages and performance. Model development will be constantly improved as extensive monitoring and adaptive management principles will also be employed to improve the knowledge base and reduce scientific uncertainty to improve the CLEAR ecosystem model and its ability to predict outcomes of planned restoration measures.

Co-Located Team

Traditionally, a USACE study will consist of a study team, known as a *Project Delivery Team* (PDT), primarily of multi-disciplinary USACE employees. During the study the PDT works with the various local, state, and Federal agencies and with the various stakeholders and sponsors of the study. However, these other groups often have only limited input to the study before the development of initial plans or once a draft report is put out for public review and comment. While this may result in the development of a good plan, one that will meet the study objectives and still address affected stakeholders' concerns, this process is usually time-consuming, often requiring significant modifications to the proposed plan to address issues not fully analyzed during the study.

It quickly became apparent that neither the USACE nor the Louisiana DNR had all the required knowledge and expertise necessary to develop a comprehensive coastal restoration plan that would meet the objectives of the LCA study. Clearly, the combined knowledge of other Federal, state, and local agencies would be necessary. The LCA PDT sought the expertise and knowledge of these agencies and other organizations that were active in defining the coastal loss problem, its causes, and potential solutions. The PDT established a co-located Team for LCA in which individual representatives of many of the other agencies literally worked fulltime or part-time on LCA at the New Orleans District offices for the duration of the study. The intent of the co-located team was to use an interagency team to evaluate proposals and work on issues directly and efficiently. The goal was to improve communication among agencies and groups, streamlining the normal bureaucratic channels to gain feedback, concurrence, and/or objections to the direction of the study and individual proposed features in a timelier manner. In addition, the best available knowledge, science, and technology could be employed to develop the comprehensive plan. Representatives of the USACE, La DNR, La Department of Wildlife and Fisheries (La W&F), U.S. Environmental Protection Agency (EPA), the U. S. Geological Survey (USGS), the U.S. Fish & Wildlife Service (USF&W), and the National Resources Conservation Service (NRCS) participated on the co-located team.

The co-located team provided enormous advantages to the study and development of the recommended plan. First, the team was able to pull from the knowledge and abilities of a staff with much more diverse capabilities than normally applied to conduct the study. Second, many issues between agencies which in the past would require weeks if not months to resolve were usually resolved in a few days unless they required senior management to resolve. Even then, by having members of those agencies on the PDT, issues were normally resolved more quickly and effectively. Third, agency members became more familiar with the practices, concerns, and policies of each other's respective agencies and developed strong working relationships. This agency networking had additional benefits in that other agencies' employees (not working on LCA) had associates working directly with representatives of other agencies. More than once co-located team members helped resolve non-LCA issues from their agencies by linking their representatives more directly with other agencies.

Co-located teams should be considered for major watershed-based studies, particularly if there are a large number of stakeholders with very divergent interests. The use of co-located teams would not be practical for relatively small studies with limited issues and stakeholders. However, the use of one or two co-located teams at USACE District offices on major studies could build stronger relationships between the various resource agencies and improve communication and cooperation on other smaller studies in the future.

Adaptive Management

Another strong aspect of the LCA plan relates to its emphasis on adopting and applying adaptive management objectives throughout the coastal zone. While adaptive management is not a new concept, its application to a large coastal ecosystem as proposed for the LCA project is a relatively new process. As previously discussed, the level of knowledge, science, and technology with respect to coastal ecosystems, while impressive, is still limited and it is not always possible to know exactly how an ecosystem will respond to multiple restoration measures. It is therefore critical to the success of the overall objectives of the LCA plan to be able to modify or adjust the restoration measures once the outputs become known and/or new knowledge, science, or techniques become available to improve the performance of the measures. Adaptive management principles are ingrained into the LCA study approach and plan recommendations.

Monitoring and adaptive management of completed projects have often been limited in costs and scope. These limits have applied to traditional projects where final designs and expected outcomes are predictable and the monitoring and adaptive management funds are necessary to verify those outcomes and adjust for minor, unexpected results. However, for the LCA project, numerous studies, designs, construction, and implementation are anticipated to be long-term processes in which initial projects will be built and operated while others are added over the project life. Monitoring of the overall functioning of the ecosystem will be required to facilitate ecosystem analysis, engineering, design, and operation of the program features. This monitoring is substantially different and cumulatively more costly than monitoring of individual projects for performance. The LCA program would dedicate more resources to monitoring provisions over the life of the LCA project to learn how the coastal wetlands are responding to the operation of new and existing projects and to help guide adaptive management as it is applied to the ecosystem.

LCA's use of adaptive management is best demonstrated by the inclusion of the Science and Technology Team for the project. The S&T program is intended to help develop and improve our knowledge and science of ecosystems' functions and how they react to various restoration measures and techniques. Also, the S&T program will help develop the technology or "toolbox" for applying that knowledge and science to coastal restoration.

LCA – After the 2005 Hurricanes

On August 29, 2005, Hurricane Katrina struck the Louisiana and Mississippi coastlines, causing devastating destruction and damage to developed areas (most notably the New Orleans metropolitan area) and the coastal marshes. A Category 3 storm at landfall, Katrina was one of the strongest hurricanes to impact the U.S. coast during the last 100 years. Less than a month after Hurricane Katrina made landfall, Hurricane Rita made landfall along the Louisiana western coast near the Texas state line on September 24, 2005. This storm caused extensive damage along almost the entire Louisiana coastline.

Hurricanes Katrina and Rita, in addition to the damage they caused to developed areas of the coastline, also caused significant damage to LA's coastal marshes. Current estimates of the marsh land lost as a result of the hurricanes vary, but the most recent USGS estimate (Barras 2006) indicates that almost 220 square miles were lost during these two storms. Since the LCA study has a projected estimate of future loss of 500 square miles by 2050, this represents a loss that exceeds projected wetland losses for the next 20 years or more. While the LCA study's 50-year estimate included anticipated losses from hurricanes, that estimate did not adequately envision the extent of the damage from Category 3 and more extreme hurricanes. It is also important to note that approximately 40 percent or more of the actual land loss from these two hurricanes occurred in areas that were experiencing low land loss rates or were actually accruing land from either natural or man-made influences. Further research is needed to gain a clearer and more thorough understanding of the role hurricanes play in coastal land loss and to help coastal restoration measures become more effective. This should result in improving scientists' ability to forecast the probable range of wetland loss that would occur over the next 50 years. While it is impossible to accurately predict the number of future hurricanes in the next 50 years, much less their respective strengths and routes, it is important to better understand their impacts on existing wetlands to enable coastal restoration designers to provide better protection for the wetlands. In particular, designing better protection for the barrier islands (or rebuilding those that have been destroyed) has the potential to slow and possibly halt wetland loss in many areas of the coastline. The proposed Science and Technology Team for the LCA project can and should provide for focused research in this area.

In response to the devastation caused by Hurricanes Katrina and Rita, the US Congress and the State of Louisiana directed their respective agencies to analyze and recommend comprehensive plans designed to provide hurricane protection up to Category 5 hurricanes while addressing coastal restoration needs. The Louisiana Coastal Protection and Restoration (La CPR) study began in January 2006. In July 2006, the La CPR Preliminary Technical Report (USACE 2006) was submitted to Congress. The report identified several potential alternative alignments for hurricane protection. Potential measures not only included hurricane protection structures and levees, but also coastal restoration features (marsh creation) designed to reduce approaching storm surge heights and non-structural measures such as storm-resistant buildings and raising of structures. The level of analysis in this report and the short schedule was not adequate to develop realistic cost estimates of the proposed measures nor quantify the benefits that would be provided. That analysis will be conducted for the 24-month final report due in December 2007. While no cost estimates are available, few involved with the study doubt that a final comprehensive plan meeting the study objectives will require a multi-billion-dollar investment.

The 2005 hurricanes have changed the coastal area priorities. The primary LCA objective was to develop a comprehensive plan for coastal restoration and preservation of wetlands for primarily environmental goals, although it noted that healthy sustainable wetlands could also provide an undetermined level of hurricane protection. While the La CPR study authority calls for coastal protection and restoration, this generally must be related to providing hurricane protection for the developed areas or providing surge reduction or buffering for hurricane protection features. This is not an insignificant distinction. Priorities in the region now are focused on hurricane protection and recovery with coastal restoration that supports that protection. Political and public support still exists for the LCA objectives but, clearly, the primary attention in the region is on hurricane protection and rebuilding. Current priorities and decisions about how LCA and La CPR overlap and support each other will determine how much momentum can be achieved for the LCA project. However these priorities and decisions are set, restoration of coastal wetlands will likely be a high-priority item. The S&T program, with demonstration projects along with ecosystem modeling, could clearly have a role in determining potential projects. Adaptive management will also be critical to the overall success of these projects. Throughout the entire process, continued close coordination and use of the best available knowledge and science from the various entities involved in coastal restoration will be critical.

Conclusions

In 2004, the LCA study proposed a $2 billion plan for initiating sustainable ecosystem restoration for the entire Louisiana coastal zone. The Louisiana coastal zone has been degrading for at least the last century, as over 1,500 to 1,900 square miles have been lost to erosion, salinity intrusion, and other natural and man-made factors. If corrective action is not taken, the system will continue to degrade, with ecosystem collapse a near certainty. While acknowledged as only a beginning, the LCA plan provided for near-term restoration measures that will slow or halt further degradation in critical wetland areas while major long-term measures are identified and studied to determine if they are suitable for the ecosystem.

Technology played a central role in the LCA study and in the development of the restoration plan for the Louisiana coastal ecosystem. From the beginning of the study, La DNR and the USACE recognized that knowledge gaps existed with respect to coastal wetland ecosystems; their interactions and linkages with the hydrologic, hydraulic, and biological processes; and how the wide range of proposed restoration measures would impact these ecosystems. The study team also recognized that knowledge and expertise existed in other agencies and groups which were needed to apply the best available knowledge and science in developing the restoration plan. The use of a co-located team to bring that expertise to the study itself greatly improved the plan developed and addressed many issues quickly during the study that, historically, would have been raised and addressed during agency and public review and comment. An additional benefit of the co-located team was the building of professional working associations among the study team members of the agencies. This has provided and will continue to provide benefits for smaller studies in which the co-located team concept would not be practical.

The LCA plan also provided for continuing to expand the available knowledge, science, and technology necessary to analyze coastal wetlands ecosystems and their processes and linkages to determine the best restoration measures to apply. The S&T program, including demonstration projects, will provide a mechanism and funding to identify the uncertainties and knowledge needed to address the coastal ecosystem problems and to develop the science and technological tools necessary. The CLEAR model developed during the LCA study is a good example of the types of tools that the S&T program could support. In subsequent phases of the LCA program, the CLEAR model will be able to take advantage of the increased knowledge and science that the S&T program will help develop, and this will allow the model to be further refined and updated. The LCA plan also called for extensive monitoring and adaptive management to determine how restoration measures are performing and to be able to make changes or adjustments to these projects to improve on the performance of the overall plan. Knowledge gained from the monitoring and adaptive management will be available to the S&T program to apply to its investigations.

The technology applied in the LCA study and restoration plan will help address the Louisiana coastal wetlands' ecosystem problems and help develop restoration measures that will apply the best available knowledge, science, and technology to those problems. The technology and tools gained will also be applicable to future restoration efforts, not only in Louisiana but in other coastal areas.

Note: The views expressed in this case study are those of the author and do not necessarily represent those of the U. S. Army Corps of Engineers, the Army, or the United States.

References

Barras, J.A. (2006). "Land Area Changes in Coastal Louisiana After the 2005 Hurricanes: A Series of Three Maps," *National Wetlands Research Center*, U. S. Geological Survey Open-File Report 2006-1274.

Boesch, D.F., Josselyn, M.N., Mehta, A.J., Morris, J.T., Nuttle, W.K., Simenstad, C.A., and Swift, D.J. (1994). "Scientific Assessment of Coastal Wetland Loss, Restoration and Management in Louisiana," *Journal of Coastal Research*, Special Issue No. 20.

Gagliano, S.M. (1972). Environmental atlas and multi-use management plan for south-central Louisiana. *Center for Wetland Resources*, Louisiana State University, Baton Rouge.

Louisiana Coastal Wetlands Conservation and Restoration Task Force (CWCRTF) and the Wetlands Conservation and Restoration Authority. (1998). "Coast 2050: Toward a Sustainable Coastal Louisiana," Louisiana Department of Natural Resources.

U. S. Army Corps of Engineers, New Orleans District, (November 2004), "Louisiana Coastal Area (LCA), Louisiana – Ecosystem Restoration Study."

U. S. Army Corps of Engineers, New Orleans District, (July 2006), "2006 - Louisiana Coastal Protection and Restoration – Preliminary Technical Report to United States Congress."

CHAPTER 5

CASE STUDIES ON EMERGING TECHNOLOGIES

This final grouping of case studies will focus on emerging technologies and how they are being applied to solve water resources challenges throughout the world. Many of the following case studies were built upon lessons learned during the two previous groupings of case studies—as well as countless others that were not included in this book. As such, this final grouping of case studies represents some of the most complex types of technologies being employed by water resources experts today and, most likely, in the near future.

South Platte River Decision Support System

Introduction

"Here is a land where life is written in water" is the first line in Thomas Hornsby Ferril's poem in the rotunda of Colorado's state capitol (Grigg 2003). The dependence on water highlighted by this famous poem has been true since Colorado's earliest settlers arrived, and now some five million residents depend on Colorado's water lifeline. Although early explorers like Zebulon Pike thought Colorado was a desert not suitable for settlement, they did not foresee the ingenuity of settlers who understood that by capturing water and storing it for later use, more land could be developed and irrigated. Now, a vast and vital water network sustains the state's economy and life.

Competition for water started before the state's Constitution was framed in 1876, which established its prior appropriation doctrine of water law. In the space of a few decades, most East Slope water was captured, and cities and farmers went after West Slope water. Later, thousands of wells were drilled into valley floors and on the plains and today's complex system of stored mountain water, trans-mountain diversions, and well water had taken shape. After the nation's environmental laws were superimposed on the prior appropriation doctrine, the amount of information required to manage this system and coordinate among stakeholders increased again. Today, any water decision involves multiple issues and stakeholders and requires decision makers to consider large amounts of information and analysis.

As a "headwaters state" Colorado's rivers flow out of the state toward the Gulf of Mexico and the Gulf of California. This mid-continent setting creates water interdependence with other states and the federal government. Decisions about in-state water use affect water deliveries to other states and vice-versa. This also creates a strong and broad need for shared knowledge from data bases, models, and cooperative planning.

A good way to provide shared knowledge for water managers, whether for state or interstate issues, is through decision support systems (DSS) and their computer-based tools. This case study focuses on use of DSS for managing water in Colorado's most populated basin, the South Platte. It describes development and use of DSS to serve the unique purposes of water management in a dry, mid-continent state. To adequately relay this case study, it is necessary to describe the management controls on Colorado's water and how the DSS developed.

Colorado's Water

Colorado is divided by its river basins, with the Continental Divide forming a major border between those flowing to the east or to the west. The regions served by these rivers can seem like different states, ranging from those like Kansas on the east to those like Utah on the west. The South Platte provides water to the populous Front Range and to an extensive agricultural economy in Eastern Colorado. The Arkansas and Rio Grande provide irrigation water in their valleys, and the Colorado and its tributaries serve diverse development, agriculture, recreation, and ecological needs on the Western Slope. Figure 20 shows Colorado's major river basins (Grigg 2003).

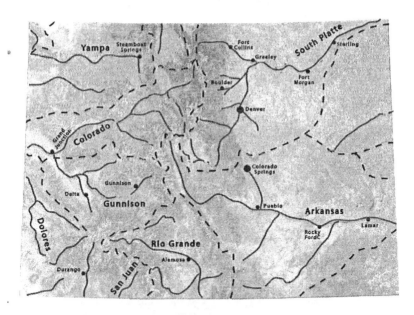

Figure 20. Colorado's River Basins.

Early in Colorado's history it became clear that most settlement would occur in the drier East Slope areas, rather than in the mountains or the West Slope. This settlement pattern created a "pull" effect on water, giving rise to an extensive system of interbasin water transfers. With today's information-based economy, more settlement is occurring on the West Slope, but not to rival the populous Front Range corridor. Figure 21 shows a conceptual view of this transfer, along with the inflows.

79

Groundwater is an important part of Colorado's water resources, and it is used for city water supply, industries, and irrigation. While most of Colorado's surface water is on the Western Slope, most of the groundwater lies east of the Continental Divide. Aquifers yielding the most water are alluvial sand and gravel deposits. These include the Ogallala, the river valley aquifers, and the valley–fill aquifers of the San Luis Valley. Eastern Colorado's High Plains are divided by alluvial strips or piedmont regions that lie along the South Platte and Arkansas valleys.

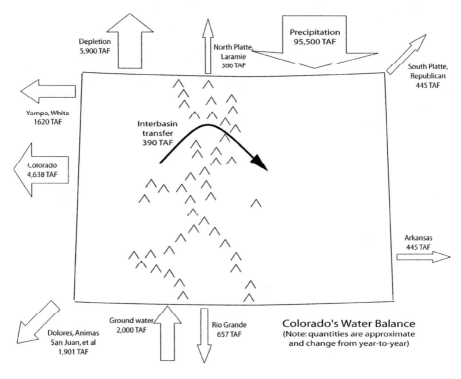

Figure 21. A Water Balance for Colorado.

In the South Platte valley the land rises from the lowest point to a series of terraces formed as floodplains. The alluvial aquifer systems in the South Platte have come under scrutiny due to the high demands placed on them. Intensive urban and agricultural development place more demands on the aquifers than they can meet, and new wells are restricted. Measures are in place to replace well water depletions with surface water rights that are in priority. Water quality in the South Platte alluvium varies mainly due to return-water quality. Dissolved solids and nutrients increase as the water is used and reused for irrigation.

Deep aquifers also supply water for Colorado needs, particularly for the growing demands in the Denver Region. Deep aquifers and non–tributary aquifers in Colorado

80

that are managed are called "designated ground water basins." The major aquifers used in the South Platte basin are the alluvial aquifers and the deeper, bedrock aquifers of the Denver Basin. The Denver Basin aquifer system comprises four aquifers under about 7,000 square miles of an area bounded by Greeley, Colorado Springs, Denver, and Limon.

Use and management of water in the alluvium requires a closely scrutinized water balance that includes recharge from precipitation and seepage from irrigated fields, from canals and reservoirs, and from other nearby aquifers. These are the hydrologic issues that are addressed in the stream aquifer models used in DSS.

The 1923 South Platte Compact joins Colorado, Nebraska, and the Federal Government. Colorado has the right to fully use the South Platte's water between October 15 and April 1. Between April 1 and October 15, if the flow in the South Platte drops below 120 cubic feet per second (cfs), use of certain water rights in the Lower South Platte may be curtailed.

Colorado's System of Water Administration

The context for use of DSS in Colorado water management is the state's system of water administration, which stems from its water laws (Grigg 1996). The Colorado system is court-based, whereas most other states have administrative systems. In a court-based system, judges hear evidence about water claims and transfers and issue decrees. In an administrative system, officials such as state engineers or boards normally decide these matters. The decisions are similar, but the one making them is different.

Colorado water courts are state district courts. Their decisions are judicial, but most rulings are by "referees" who work for the water judges. After decrees are issued, it is up to the State Engineer to administer them on a daily basis. To do this, the State Engineer's Division of Water Resources operates regional offices in water divisions in the seven river basins of the state. Each of these is headed by a Division Engineer who oversees a group of water commissioners who work with local water users to administer the water diversions and records. The Water Commissioners actually make the on-the-ground decisions about who gets water and when.

Surface water rights are prioritized according to the dates of the decree. A water user might have an 1890 direct flow right to 5 cfs, for example. That would mean that the user could divert up to 5 cfs, in priority, during times when he or she had historically used the water. However, the user could not change the time, place, rate, or schedule of use without going to water court. It is up to the State Engineer's forces to determine when the water user can divert.

Water users may be senior or junior, depending on the level of priority. A junior upstream user may have to forgo diversion so that the water will travel downstream to meet a senior's needs, for example. Colorado operates on a system of "calls." A call occurs when the Division Engineer determines that there is enough water in the river to meet rights to a certain date, say 1910 rights; then all juniors must stop diverting. Determining when a call is necessary is the main technical challenge facing the Division Engineer. If the determination is not right, water may be wasted or a downstream user may not get water.

Groundwater was integrated into the surface water system in 1969. Now, to pump wells, the user has to be in priority. To provide for the users who drilled wells before 1969 but at junior dates, plans of augmentation were included in the law. These enable users to provide replacement water at times when the call is on the river so that the well can continue in operation. It has proved difficult for well owners to find this replacement water during dry periods, and this problem has led to a recent shutting down of well pumping, causing great hardship in the South Platte basin.

Theoretically, Colorado's system looks neat, but working out problems in the face of differing needs and imperfect information is a challenge. One water attorney said that there are three ways that one can view the system of administration: the way it is supposed to work, the way people think it works, and the way it really works. What this means is that deals, trades, special arrangements, and other compromises must inevitably take place among local water users. The role of the Water Commissioner is thus to serve as coordinator and facilitator as well as administrator.

In Colorado's system, decisions are made according to users in an entire river basin. However, the lack of perfect information has been a deficiency of this method and illustrates the need for a decision support system.

The South Platte River Basin

From its mountain headwaters near 14,000 feet, the South Platte flows toward Denver and the plains of Eastern Colorado, dropping to an elevation of 3,500 feet at Julesburg. At its North Platte confluence in Nebraska, it drains 24,300 square miles, including 19,020 square miles in Colorado. To easterners, the South Platte may seem more like a creek than a river, but it provides essential water for Colorado's cities and agricultural economy. Water in the basin comes from direct surface runoff, from interbasin transfers, and from groundwater. The South Platte can also experience large floods, as it demonstrated in 1965. It is a water management corridor that provides for multiple uses from mountain environmental uses to conveying water to cities and farmers downstream (Colorado Water Resources Research Institute 1990).

The South Platte has been modified extensively since its first developed uses before the 1859 gold rush. Before development, it was a mountain-to-plains stream that ran low during the fall and winter months and then flooded with the annual "spring runoff," which scoured and maintained its wide flow path. The first diversion of water from its basin was in 1859 and by 1909 some 1.1 million acres were under irrigation.

The South Platte draws more imported water than any other basin in Colorado and is the artery serving many communities and farms from the mountains to the state line. It hosts the extensive and water supply systems for Denver and its suburbs, as well as many irrigation companies and water districts.

Precipitation in the South Platte basin varies from about 12 inches per year in the plains to over 50 inches in the mountain headwaters. About 70% of the basin's water supply is from snowpack, which runs off in the spring. Annual runoff in the basin (not including imports) is about 1.9 million acre-feet, but it varies widely from dry to wet years. Flows leaving the state also vary, averaging about 350,000 acre-feet. Water imports to the South Platte from some 20 tunnels and ditches bring about 370,000 acre-feet annually. The South Platte's alluvial aquifer is used extensively and has some 25 million acre-feet of storage. Deep aquifers have much more storage and a low rate of recharge.

Today, there are approximately 4,500 direct flow and 1,300 storage rights in the basin, or 6,200 water rights. Some 370 reservoirs can store about two million acre-feet of water, and some 542 irrigation diversions water about 1.2 million acres of land. Denver and five other major metropolitan areas depend on the South Platte for water. Because the South Platte's water diversions and use exceed its supply, water is used more than once. That is, one user's return flow becomes another's supply.

The search for new water supplies to feed growing populations dominates the news in the South Platte basin. For example, in the 1980s the Denver suburb Thornton secretly reached northward to buy farms and obtain water rights in Larimer and Weld counties. Another controversy surrounded the failed 1980s proposal to build Two Forks Reservoir on the South Platte above Denver. The South Platte also attracts controversies over use of groundwater. One of these is use of the Denver Basin aquifers, a series of deep formations underlying Denver, which have become important for water supply for growing communities, particularly in Douglas County. More recently a large number of South Platte Basin farmers were cut off from their well supplies by a court ruling.

The South Platte and its urban tributaries also form recreational corridors through the Denver area. After the floods of 1965, local agencies created improved channels and bike trails and kayak chutes. Today, the river is the centerpiece of an attractive and restored South Platte valley through Denver.

The South Platte River joins the North Platte River at North Platte, Nebraska to form the Platte River, which flows through Nebraska and joins the Missouri just south of Omaha. Three states— Colorado, Wyoming, and Nebraska—are involved in the Platte River Basin and decisions about the South Platte affect all three.

The Platte River is also Nebraska's water lifeline. From its natural form as a wide, shallow, braided stream, it has decreased in width due to reduced flows and sediment loads and growth of woody riparian vegetation. These changes have affected six endangered species, and in response a comprehensive management plan has been developed to involve the three states and federal government.

DSS for Colorado Water Management

Given its dependence on water and the unique nature of its court-based water rights system, Colorado has little slack in managing water. Therefore, use of DSS to level the field and create a basis for fair decisions is logical for the state. Over the years the State Engineer's office became the repository for water use records and engineers and attorneys had to go there and search manually through data to analyze proposals for water management changes (Grigg 1996).

After computers were developed, it was natural that hydrologic models would be used to analyze water issues, and today their use is widespread. One example of a river basin model is MODSIM, developed from the 1980s by John Labadie at Colorado State University. It is used by cities and water districts for water accounting. In 1991, John Eckhardt, then of the State Engineer's Office, developed a DSS for operating reservoirs under the appropriation doctrine. The problem addressed was how to operate reservoirs with water belonging to different water right owners under a complex system of natural water and stored water that resulted from the unique features of the appropriation doctrine. Eckhardt's framework included an operator interface, a provision for system simulation, and an information-management subsystem. With the need to analyze reservoir releases, surface water flows, and groundwater effects, it was inevitable that stream-aquifer models would enter the picture. A surface-groundwater hydrologic model named SAMSON, for "Stream-Aquifer Simulation Model," was developed by Hubert Morel-Seytoux at Colorado State in the 1980s. These three examples of models illustrate the kinds of technologies that have merged into the current development of water DSSs that can be used in Colorado water-management decisions.

The Colorado River DSS

During the 1990s, Colorado's state water agencies decided that a computer-based decision support system would help them work with the Bureau of Reclamation and other states to manage the Colorado River. This led to the Colorado River System Decision Support System (CRDSS), which was the first DSS developed by these agencies (Dames and Moore and CADSWES 1993).

The project was to be a cooperative effort between the Department of Natural Resources, the Colorado Water Conservation Board, the Division of Water Resources, water users, and other interests. A bill was passed in 1992 to provide funds for the DSS. A feasibility study was completed that identified three purposes of the DSS:

- Interstate compact policy analysis, including evaluation of alternative operating strategies, determination of available water remaining for development, and maximization of Colorado's compact apportion.
- Colorado water resources management, including development of basin-wide planning models, examination of water management options, and evaluation of impacts of instream flow appropriations for endangered species.
- Colorado water rights administration, including optimization of water rights administration in Colorado, on-line sharing of information between water users, and the potential administration of water rights under a compact call.

A four-year development period for the DSS began, with initial efforts focused on data base design and construction, development of the ability to use the Bureau of Reclamation's Colorado River Simulation Model, the development of a consumptive-use model, and assembly of a water- rights planning model.

SPDSS Feasibility Study

After the CRDSS process was well underway, the State Government decided to create decision support systems for each of the other major river basins in Colorado, including the South Platte. The feasibility study for the South Platte Decision Support System (SPDSS) was completed in 2001, some eight years after the pilot effort on the CRDSS (Brown and Caldwell et al. 2001). The SPDSS was to encompass the South Platte and North Platte River basins and to use data to characterize the hydrologic and hydrogeologic features of the basins and tools for water administration and planning. The other DSS that is nearing completion is on the Rio Grande basin (RGDSS) with rules and regulations for new well development in the San Luis Valley and analysis of its Closed Basin Project and Interstate Compact operations.

The initial phase would focus on basic water resources data before significant modeling occurred. It would allow use of data in water administration activities to be implemented quickly, to be followed by later models. After the initial phase, the SPDSS was to focus on water budgets, consumptive use, a groundwater model for the Denver Basin, and surface water models. The last phase would include a groundwater model for the Lower South Platte alluvium.

Experience with models and the CRDSS shows that adequate attention must be given to funding of DSS development, operations, and maintenance. Development cost for the recommended version of the SPDSS was $15 million (38% for data collection, 46% for components, and 16% for other tasks). This was to be financed from the CWCB Construction Fund rather than current tax revenues. The SPDSS was projected to cost some $420,000 per year to be operated by three employees to handle models, GIS, and databases. Operational expenses were to be split between the general fund and the CWCB Construction Fund, with plans to shift expenses to the general fund later.

The SPDSS was to be developed over a six-year period in three phases. The first phase would focus on data collection and water administration tools. An existing "StateMod" model would be enhanced to include river call reporting by node and time step and to incorporate updated data. Later, an enhanced groundwater model for the Denver Basin region would operate on a monthly time step and use a one-quarter to one-square mile grid.

Some of the features of the SPDSS would include a groundwater model of the Lower South Platte alluvium on a monthly time step; access to historic priority call data via the Internet and CDs; new methodologies to calculate consumptive use, such as the Kimberly Penman method; access to satellite images and maps of irrigated lands; access to a GIS network of surface water hydrology, structures, and water distribution systems; and graphical and visual displays of model results.

The DSS should increase the ability of the State Engineer to administer water. Access to better information about the state-of-the-river was to aid water accounting. Features to help with this are automated call notification; improved access to real-time streamflow and diversion data; improved to allow direct entry of data by users into a real-time database; better access to historic river call, augmentation plan, substitute supply plan and transfer decree data; improved ability to analysis of real-time or historic data; and access to animation tools for presentation and visualization.

Planning tools were to provide for evaluating success of species management, improving applicability of streamflow depletion factors for groundwater analysis, determining effects of groundwater pumping, and providing dataset quality assurance/quality control. To achieve these, the DSS would include:

- A basin-wide water resource planning model (StateMod) to operate on a monthly or daily time step and include all of a basin's consumptive use;
- Training programs to facilitate access to data by water users;
- Efforts to locate non-exempt wells;
- Addition of monitoring wells to provide additional geologic structure, aquifer property, and water level data;
- Field studies to characterize streambed conductance;
- Aquifer tests using existing pumping wells;
- Estimation of municipal well pumping based on user interviews, population data, and water use data;
- Estimation of irrigation pumping;
- Access to improved consumptive use and irrigated acreage data;
- Access to transit loss data;
- Access to well location, water level, and pumping data;
- Expansion of stream depletion factors into tributary areas where they do not presently exist;
- A GIS database;
- Tools to access data and models;

- Access to crop coefficients for importation to the South Platte Mapping and Analysis Program; and,
- Mapping of land use, diversion structures, and irrigation distribution systems with water use linked to the irrigated acreage.

Conclusions and Lessons Learned

The development and implementation of the state's water DSS is an ongoing process and more time is required to report all of their lessons. Based on some experience of the writer with creation of the CRDSS and with research about South Platte water management, the following are offered as some early observations.

The DSS are timely and needed because water management continues to grow more complex. Water management in Colorado has legal implications and sometimes involves high financial stakes and interstate issues. Thus the DSS have political overtones. Given the lack of slack in the state's water management systems and the legal, financial, and political implications of water management decisions, it is imperative to lend transparency and aid cooperation by providing valid and relevant data. That the state legislature would finance the DSS is evidence that political support exists for this view.

The DSS serve three levels of government and the private sector. Their work spans planning and operational functions. Creating and validating them is a good way to promote intergovernmental coordination and shared vision planning.

Creation of Colorado's DSS is a logical extension of the key role of state water agencies in coordinating water management in the state. The DSS address primarily water quantity and use, topics that are driven mainly by state water law. Thus, professionals in state government have a large part in designing and operating these systems. They have been supported by talented consulting firms and individuals who were available to create the systems. These firms included individuals with advanced degrees and/or years of experience in data management and modeling.

Water quantity management in Colorado is complex due to the extensive data and analysis required on rivers and aquifers where water supplies and demands change continually and rights of use depend on a system of priority. Thus, the DSS must be designed to provide the required comprehensive information as do enterprise DSS systems in organizations. That is, in addition to basic hydrologic analysis, the systems must be able to retrieve information on historic use and diversions and to align this information with current conditions so that calls can be made and enforced.

The DSSs require significant financial support to develop and operate, and general fund appropriations are necessary to support them. Later there will be the inevitable calls to have them financed from "user fees," and the sustainability of the DSS will depend on the resolution of these financing issues. Large quantities of data are used in the DSS and managing the data bases will continue to be a complex undertaking.

Also, the models will never become routine, involving storage-routing and surface and groundwater models, with emphasis on stream aquifer models. Thus, the DSSs will require significant continuing efforts to manage and validate.

References

Brown and Caldwell, Camp Dresser & McKee, Bishop Brogden Associates, and Riverside Technology. (2001). "South Platte Decision Support System Feasibility Study". Colorado Water Conservation Board and Division of Water Resources. Denver. October.

Grigg, N. (2003). *Colorado's Water: Science & Management, History and Politics.* Aquamedia Publications. Fort Collins, CO.

Grigg, N. (1996). *Water Resources Management: Principles, Regulations and Cases.* McGraw-Hill. New York.

Dames and Moore and CADSWES. (1993). "Feasibility Study, Colorado River Decision Support System". Colorado Water Conservation Board and Division of Water Resources. Denver. Jan 8.

Colorado Water Resources Research Institute. (1990). "South Platte River System in Colorado. Working Paper". January, 1990.

Colorado Division of Water Resources. Web page on South Platte Decision Support System. http://cdss.state.co.us/. (January 16, 2007).

Lake Ontario-St. Lawrence River Study

Introduction

Lake Ontario is the most downstream of the Great Lakes that define part of the boundary between Canada and the United States. It receives its water from the four other Great Lakes as well as from its local watershed. It discharges water into the St. Lawrence River that flows northeastward past Montreal and Quebec into the Gulf of St. Lawrence and the Atlantic Ocean. The levels of Lake Ontario and the upper portion of the river and the flows and levels in the lower portion of the St. Lawrence River are regulated, to some extent, by the operation of the Moses Saunders Dam that separates these two river portions, as shown in Figure 22.

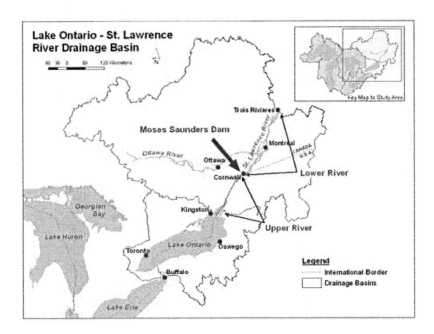

Figure 22. Map of Lake Ontario and St. Lawrence River and their watersheds.

Water Level and Flow Regulation

The International Joint Commission (IJC), which oversees all transboundary waters along the entire Canadian-US border, established criteria for the operation of the Lake Ontario-St. Lawrence River system and created a Board of Control to develop and implement plans of regulation to meet the criteria. In 1963 a new plan called "1958D" was approved and implemented. The plan consists of rules for making releases through the dam every week based on how high the Lake is, the time of year, ice conditions, Ottawa River flows, river stages, and a series of flow limits.

The dynamics of regulation for the Lake Ontario-St. Lawrence River system are complex, requiring the balancing of several conflicting water management objectives that are inherent in the management of flows and lake levels. For example, alleviating high water levels on Lake Ontario requires releasing more water, which may cause flood-related damage downstream because of high water conditions in the River. Alleviating low water levels in the lower river requires releasing more water from Lake Ontario, which may cause problems for recreational boaters and municipal water suppliers along the upper river and lake shore because of low water levels. Managing the variability of water levels to accommodate ecosystem needs introduces a higher level of complexity.

Uncertainty about future water supplies from the Upper Great Lakes and tributaries within the Lake Ontario-St. Lawrence River Basin makes it that much harder to know how much to release through the dam to balance upstream and downstream needs. For example, if future supplies are unexpectedly low, releases made to alleviate low water levels in the River may drain too much water from Lake Ontario, making it much more difficult to alleviate low river levels later when the impacts may be even worse. Similarly, if future supplies are unexpectedly high, restraint in making releases to avoid minor downstream flooding may induce greater damage when later releases have to be increased dramatically.

The timing of water availability within the year is important, in different ways for different purposes. The level of commercial navigation and recreational boating activity drops considerably in the winter. The value of energy generated in the summer during peak energy demand periods can be more than 12 times the value in the spring. Higher releases reduce the level of Lake St. Lawrence, which is immediately upstream of the hydropower dam. If the releases are too high, the levels can be so low that they are hazardous to navigation and could result in ship groundings. In addition, high flows can produce cross-currents that make it difficult to control vessels. For hydropower, more electricity can be generated when a greater volume of water passes through the turbines but this reduces the head, the level of water in Lake St. Lawrence upstream of the dam, and this in turn reduces the amount of electricity generated for each cubic meter of water.

The operation of the lake levels and river flows require that riparian interests downstream receive no less protection from flooding than would have occurred under pre-hydropower project conditions. In other words, the release operating policy can not make conditions worse for shore line owners than what would have occurred before the construction and operation of the dam. Regulation of Lake Ontario outflows has actually reduced spring flooding in the Montreal area, while still reducing flooding on Lake Ontario. Montreal is threatened by flooding since it is located at the confluence of the Ottawa and St. Lawrence rivers. The spring runoff from the Ottawa River Basin is largely uncontrolled and can be very significant. Timely adjustment of the Lake Ontario outflow has repeatedly helped avoid serious flooding around Lake St. Louis in the Montreal area during Ottawa River floods. Lake Ontario outflow reductions are typically offset by higher flows before the Ottawa River flood or shortly after it.

The regulation of Lake Ontario water levels and outflows began in 1960. The current plan, 1958D, has been in operation since October 1963. The Plan's authors designed it for the hydrologic conditions experienced from 1860 to 1954. For that reason, 1958D has not performed well during the extreme high and low water supply conditions experienced since then. As a result, the International Joint Commission and its Board of Control have had to deviate from their 1958D Plan. More recently, the Board of Control has deviated from the Plan to better address changing needs and interests, mainly recreational boating and environmental and ecological issues.

A Policy Review

In April 1999, the International Joint Commission requested the Governments of Canada and the US to fund a review of the regulation of Lake Ontario levels and outflows in light of public concerns and in response to potential climate change conditions. They agreed and 20 million dollars and five years were allocated to this review. An International Lake Ontario-St. Lawrence River Study Board was created, with seven professionals from each country.

In December, 2000, the Commission issued a directive to the newly created Study Board to:

• Review the current regulation of levels and flows in the Lake Ontario-St. Lawrence River system, taking into account the impact of regulation on affected interests.

• Develop an improved understanding of the system by all concerned.

• Provide all the relevant technical and other information needed for the review.

To carry out these tasks the Study Board created various technical working groups, each associated with a particular water impact, use, or interest group and data need These technical working groups enlisted the help of consultants when necessary. The Board and technical working groups met periodically at various sites within the basins. A major effort was needed just to keep up with every activity that was taking place and to keep everyone involved focused on the need of those creating operation policies as functions of water levels and flows.

Five years and 20 million US dollars later the Study Board submitted to the IJC its final report (LOSLSB 2006). It summarized findings from the numerous scientific studies performed, and documents prepared, for the Study. The Final Report offers three new candidate plans to the Commission and presents recommendations on public involvement and on changes in how the Board of Control operates related to the implementation of any new plan.

Over the five-year study period, hundreds of people and dozens of organizations participated directly in the Study. A Public Information Advisory Group was created at the same time the Study Board was created to enhance communication between the Study Board and public interest groups and any interested stakeholders. The volunteers of the Public Information Advisory Group (PIAG) have been central to the success of the undertaking, contributing significantly and uniquely to the work of the Study Board. Everyone recognized from the start that unless the public supports any proposed plan, the plan cannot succeed, and to support any plan, the public needs to understand it and have a role in its development. Via PIAG, the Study Board obtained public input during the development of all candidate plans.

This study represented a unique opportunity to make a change – to see if the overall operation of the system could be improved and if so to determine how. In the opinion of most of the study board members, this undertaking has succeeded in developing three candidate plans that perform better than the current operating regime in terms of overall net economic and environmental benefits to the various interests throughout the system. The Study Board is confident that each of these three candidate plans will satisfy most of the affected interest groups. Tradeoffs among the competing interests exist within any plan and among the three plans. The Study Board has identified and quantified these tradeoffs to the extent possible.

The Study Board did not rank or prioritize these three candidate plans. This job, that of determining which tradeoffs are best, is the job of the International Joint Commission.

Plan Formulation and Evaluation

Over the five-year study period, the Study Board and its consultants collected considerable physical, economic, environmental, and ecological data and performed numerous scientific analyses. Much of this required the development and use of computer optimization and simulation models, for example to assess the economic, environmental, and ecological impacts associated with particular operating policies. The Study Board also used these models to estimate the impact of possible climate induced changes in regional hydrology. New findings, conclusions, and clarifications of previously uncertain views and theories were developed during this work.

The models developed and used for this study included the following:

- Flood and Erosion Prediction System (FEPS) model for assessing shoreline erosion rates and the damage over time due to flooding and wave action on riparian property for Lake Ontario and the Upper St. Lawrence River. The model isolates the influence of lake levels for the purpose of comparing different regulation plans. There is some uncertainty in the frequency of the failures and thus overall magnitude of the economic damages. However, for comparing the impacts of different regulation plans on this PI, this uncertainty will not influence the results.
- The St. Lawrence River Model (SRM) estimates water level impacts on existing shoreline protection works, such as structural failures or maintenance events, and the associated economic costs.
- Integrated Ecological Response Model (IERM) for estimating how different regulation plans may impact the ecosystems in Lake Ontario and the St. Lawrence River. Various submodels of IERM focused on the responses of different vegetative and animal species to varying water levels.
- Various policy generation models using stochastic optimization as well as simulation for identifying and evaluating real-time operation policies and for operating policies that can be implemented without periodic modeling. These

numerous plans could then be simulated in more detail using both FEPS and IERM listed above and the overall shared vision model.

- Statistical hydrologic models were used to generate alternative time series of inflows to the system that were then used in policy simulations to ensure the reliability, resilience, and robustness of each policy. Some of these time series included up to 50,000 years that were used to analyze four climate change scenarios. Each of the candidate plans was thoroughly tested to ensure that none had fatal flaws that would inhibit its performance under potentially plausible extreme conditions.

- A Shared Vision Planning Model was developed for this study that combined all other model results, the environmental science, the economics, and public input into an interactive analytical framework that helped the Study Board and public interest groups explore numerous plan formulation opportunities, operating nuances, and performance impacts in an organized fashion. This could be done via the Internet.

The overall results of the shared vision simulation model were displayed in an interactive Excel-based program called *The Boardroom*. The Boardroom combined all study models and data as needed to provide a relatively quick way of exploring different alternative plans under different input assumptions. The Study Board members and stakeholders alike used The Boardroom to better understand the economic and environmental impacts of various regulation plans and compare their performance. The Internet version of the Boardroom summarizes the three candidate plans but does not allow model runs with user-defined inputs.

Shared Vision Model

The *Shared Vision Model* is the name of the computer model developed to integrate all the data generated in the study, from models as well as from other scientific studies, in one place. Using this model various regulation plans could be run through an evaluation process and the results compared between interests and locations. The Shared Vision Model connected all study research to the guidelines the Study Board developed to identify the best alternative plans. It integrated plan formulation and evaluation so that new regulation plans could be designed and immediately evaluated. The fact that specific mathematical connections had to be made between research products and the questions the Study Board wanted to address. The Shared Vision Model enabled everyone involved to understand how actions taken to affect one part of the system or one interest affect all other stakeholders.

The Shared Vision Model developed for this study is actually a pyramid of four models used to produce estimates of plan performance: the STELLA model, with dynamically linked Excel input files; the Flood and Erosion and Prediction System (FEPS); the St. Lawrence River Model (SRM); and the Integrated Ecological Response Model (IERM). The STELLA portion includes all of the system hydrology and all of the performance indicator relationships to water levels and flows for recreational boating, commercial navigation, hydropower, municipal and industrial water uses, and lower St. Lawrence River flooding. The Control Panel and Data Warehouse are Excel files that store and feed data to the other models. The Board Room is an Excel file where all plan results are presented using tables and graphs.

The "Shared Vision Model" approach tries to combine all the study information in a single computer model in such a way that decision makers and stakeholders can ask "what if" questions and get answers about how the things that are important to them are affected.

Data Management

The development of an Information Management Strategy (IMS) was deemed important by this Study Board for long-term use of data assets compiled or created within the study. Thus an Information Management Strategy was developed that included a comprehensive assessment of available information resources, likely future additional resources, capabilities of partners and alternative approaches for integrated information management, and data-access constraints and limitations. The Information Management Strategy promoted improvements in data storage, discovery, evaluation, and access, all of which were addressed by this study.

The Information Management Strategy chosen for this study focused on using the Internet for information discovery, evaluation, and access. The components of the Information Management Strategy are depicted in Figure 23.

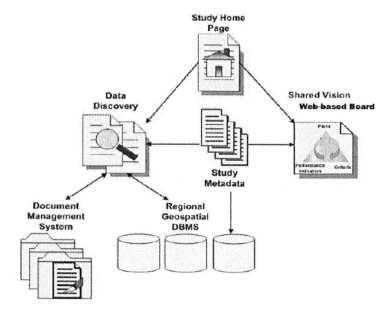

Figure 23. Schematic of Data Management System.

94

Making a Decision

From an interest perspective, all three candidate plans provide increased benefits to commercial navigation and hydropower and have no impact on municipal, industrial, and domestic water use relative to the current operating policy. The greatest difference between the three plans is in how they address recreational boating, the shoreline flood and erosion or coastal interest, and the environment or natural ecosystem.

One of the three candidate plans was designed to provide high overall net economic benefits for upstream and downstream recreational boaters and benefits for shore protection maintenance and flood concerns on Lake Ontario. Compared to the current policy, it results in higher erosion rates to unprotected Lake Ontario shoreline and higher flooding damages on the lower St. Lawrence River. It provides small improvements for the environment.

Another candidate plan strives to return the Lake Ontario-St. Lawrence system to a more natural regime similar to conditions that existed before the St. Lawrence River Hydropower Project, while attempting to minimize damages to present interests. It provides overall improvements for the natural environment on Lake Ontario and the upper St. Lawrence River compared to the current plan. Its downside is that it results in higher damages for Lake Ontario shoreline properties and some increased flood damages on the lower St. Lawrence River. In the eyes of many, it is the only candidate plan that consistently transforms and improves the diversity and productivity of the natural ecosystem, addresses Species at Risk legislation objectives, and improves ecological integrity.

The intent of the third candidate plan is to increase the net economic and environmental benefits of regulation compared to the current plan without disproportionate losses to any interests. In this respect this plan yields increased net benefits for recreational boaters and some improvement for the environment generally across the range of performance indicators considered but results in some small losses to properties on Lake Ontario.

The final decision by the International Joint Commission will be a difficult one, as it tries to balance all interests equitably. Stakeholders in the basins tend to be focused on their particular interests rather than on everyone's interests and the necessary tradeoffs among conflicting interests that will have to be made. Many interest groups know how to influence politicians in both governments, as well as within the IJC. While the final decision will be largely political, the Study Board has given the International Joint Commission a comprehensive set of tools, models, supporting data, and information that it believes can facilitate this political plan selection process.

Conclusions

For this book on planning and management technology, it seems appropriate to end this case study with a question. What happens to the technology and data developed in studies like this when the studies conclude? While the actual decision of which candidate plan, if any, will be selected has not been made at the time of this writing, the work of the Study Board, its technical working groups, and consultants is over. The Canadian and US Study Board Managers and their staff have other jobs. Unless maintained and upgraded by some agency or group of agencies, much of the technology and information developed and compiled will be lost, except perhaps as reported in scientific journals.

In an ideal world some way to maintain, improve, and update the technology – models, their data, and the data management system – would be of value to those studying the management of these waters in the future. But this requires resources, resources that can be used for other activities as well. The Study Board gave considerable thought to this issue, made some recommendations, but likely to no avail if resources and a commitment to do this are not forthcoming.

One strong argument, it seems, for maintaining some of the policy evaluation capabilities developed over these past five years is to make it easier to perform such evaluations periodically in the future when warranted. Some of the uncertainty with respect to many of the predicted impacts, especially the environmental and ecological ones, might be reduced if monitoring programs were implemented. One could determine over time just how well the selected policy is performing, and adaptive changes to the operating policy could be made if justified based on the new knowledge gained from that monitoring program.

Conditions and the priorities for lake level and flow regulation always change over time and new science and technologies will continue to evolve. So will policy goals and priorities. An adaptive management process should support the selected regulation plan incorporating performance tracking. Performance review of the new plan could be undertaken every five years after its implementation and a more in-depth evaluation could be carried out and include consideration of adaptive changes to the selected plan.

References

Lake Ontario – St. Lawrence Study Board (LOSLSB). 2006. "Options for Managing Lake Ontario and St. Lawrence River Water Levels and Flows, Final report of the International Lake Ontario", St. Lawrence River Study Board, International Joint Commission, Washington, DC and Ottawa, Canada.

Shared Vision Planning

Introduction

Technical analysis has always played a critical role in water resources planning, and the technology to support that analysis has consistently grown as technical understanding and computer modeling has blossomed over the decades. Concurrent with these changes has been an increased accessibility of the technical information and increased expectations for involvement in the analysis by a broad range of potentially affected groups. The case study in this book of the role of technology in drought planning for the Washington Metropolitan Area shows how technical tools were developed and then used with various affected parties to develop and agree to a water-resource solution.

Since the studies on water supply for the Washington Metropolitan Area in the late 1970s and early 1980s, this march toward more involvement by interested parties in the technical analysis has continued. In the National Drought study in the late 1980s and early 1990s, the Shared Vision Planning technique was born. In the decade and a half since then, Shared Vision Planning's combination of traditional planning principles, technical systems analysis, and collaboration has been applied and modified in water planning studies around the nation. Recently, Shared Vision Planning practitioners have joined forces with developers and proponents of similar methods with different nomenclature (such as Computer-Aided Negotiation, Participatory Modeling, Collaborative Modeling, Mediated Modeling) to further develop the concepts and best practices for combining the use of computer models within multi-stakeholder water resources planning processes.

This case study reviews the origins and concepts of Shared Vision Planning, describes some applications, and explores new frontiers in "the most important concept in [the water resources planning] field in the last decade and a half" (Lund 2007, personal. communication).

Origins

The water supply study for the Washington DC metropolitan area that was cited earlier in this book points out a number of key points for the success of the study:

- Citizen involvement
- Use of simulation by stakeholders to develop and evaluate alternatives
- Political buy-in
- Consideration of different objectives that included risk management
- A paradigm shift in planning process

The need and will to continue to engage affected parties in water supply planning and management for the Washington area is still in force through the annual drought exercises that take place with the various water utilities under the auspices of the Interstate Commission for the Potomac Basin and their technical models of the system.

The successes on the Potomac were the precursors to the development of a new way of water planning. And it was in the National Drought study, started in the late 1980s, where some of the ideas from Washington were further developed and applied across the nation.

Congress directed the USACE to undertake a National Drought study (IWR 1994), and the Institute for Water Resources assumed the lead of this study. The National Drought Study used multiple case studies to develop and test the "Drought Preparedness Study" method (later renamed Shared Vision Planning). The USACE had decades of solid planning principles developed from the Harvard Water project— but this was a different planning problem. Instead of the Federal government having funds in hand for a structural solution to a problem (as was typical in the 1960s, 70s, and early-to-mid 80s), USACE planners were no longer exclusively in charge. Any solution to drought issues might involve the USACE, but would also involve local governments, tribes, and other stakeholder groups. Quickly USACE planners realized the need to change the traditional planning mentality to become more collaborative and more engaging of interests in the planning and analysis process. There were three common characteristics of the drought preparedness case studies:

- Collaboration
- Reliance on but enhancement of traditional planning principles
- Integration by a technical systems model

In each of the five drought-preparedness case studies a team of stakeholders (federal, state, and local agencies, tribes, and environmental groups) collaboratively developed a simulation model of the river system to evaluate alternative actions during a drought. Using the collaboratively developed models within a planning framework, each case study developed and evaluated plans for how to operate reservoirs during drought. The planning study for the Kanawha River (WV/NC/VA) resulted in operational changes at Summersville Reservoir (WV) that preserved water quality and avoided economic losses to the white-water rafting industry during droughts (Punnett and Stiles 1993; IWR 1994). The James River (VA) model simulated drought impacts and collaboratively examined alternatives including regional management and conjunctive use of emergency wells. In the Marais de Cygnes-Osage Rivers, the collaborative modeling process improved understanding and cooperation between the states and the USACE. In the Green River study, the computer-aided process enabled stakeholders to arrive at a consensus on an appropriate refill strategy (IWR 1994).

Shared Vision Planning

Shared Vision Planning (SVP) integrates tried-and-true planning principles, systems modeling, and collaboration into a practical forum for making resource management decisions. Shared Vision Planning addresses the need for broad involvement by conducting the technical analysis collaboratively. What is different about Shared Vision Planning?

- The planning process is designed for conflicts that involve multiple decision makers.
- The collaborative, integrated, and transparent nature of the modeling sets SVP apart from traditional technical analysis.

SVP has three basic elements: (1) an updated version of the systems approach to traditional water resources planning developed during the Harvard Water Program; (2) an approach to public involvement called "Circles of Influence;" and (3) collaboratively built, integrated computer models of the system to be managed (Palmer et al. 2007). SVP uses conflict resolution methods to resolve differences that occur during planning and employs a method of collaborative decision making called "informed consent" to make decisions internally consistent, more defensible, and transparent.

In other words, SVP promotes an interest-based negotiating and decision-making environment by emphasizing the fundamental objectives of the interested stakeholders and intensively and iteratively engaging them throughout the process. In this way SVP combines traditional planning (which had not historically included intense stakeholder engagement) with collaborative processes and stakeholder involvement. By engaging the public throughout the planning process, SVP promotes collaborative learning by the stakeholders and the technical information providers, incorporates information that might have been missed otherwise, and promotes understanding and consensus building for the water resource decision. SVP relies on a traditional planning approach to protect the broad public interest and prevent undue influence by well-organized interest groups.

By linking the traditional planning approach to collaboration, SVP takes advantage of the extensive intellectual achievements of traditional planning and analysis to help prevent the capture of a public process by well-organized or vocal special interest groups. By embracing academically rigorous cost-and-benefit-analysis techniques, SVP provides decision makers with impacts to the general public that can be considered along with impacts to members of the public with special interests in the decision. SVP uses a collaboratively built systems model that fosters a common understanding of the facts.

Within SVP the collaboratively built model becomes the "single-text" negotiating tool where all the information and understanding of the workings of the system are contained in one transparent, trusted decision support tool (Delli Priscoli 1995). This tool allows stakeholders to run "what-if" scenarios to explore potential alternatives and their impacts on their objectives and the objectives of other stakeholder groups. SVP integrates the technical analysis across stakeholder interests, allowing collaborative learning about goals, objectives, constraints, and alternatives. By integrating the system interactions in one model, SVP allows the impacts on various interests to be integrated and evaluated simultaneously. As stakeholders explore options, the integrated model displays the potential impacts of alternatives on relevant issues such as recreation, ecology, flooding, and water supply reliability.

The SVP process requires transparency throughout the entire process to encourage understanding. Within an SVP process, the models need to be open and transparent so that stakeholders understand where the information and relationships come from. Such transparency helps build trust in the model and its results and also builds confidence in the collaborative process and across the team of stakeholders engaged in the SVP process.

Shared Vision Planning traces its first element, traditional planning, back to the concepts of the Harvard Water Program (Maas et al. 1962) and the 1960s era North East Water Supply Planning Study (Major and Schwarz 1990) that were later encoded in the Federal "Principles and Standards for Planning Water and Related Land Resources" (U.S. Water Resources Council 1973). During the National Drought Study, the USACE modified the P&S planning steps to make them more suitable for drought management by explicitly incorporating more collaboration and teamwork. Because drought-management decisions generally do not involve significant federal funding, these decisions typically must be agreed to by multiple management entities, with multiple levels of government controlling the drought response.

The SVP process embraces a structured approach to public involvement through the "Circles of Influence" method as developed by Robert Waldman during the National Drought Study (IWR 1994). Typically, stakeholder involvement by the federal government has focused on presenting or gathering information, which is classified as a low level of involvement (Arnstein 1969). In contrast, the Circles of Influence approach engages stakeholders in a high degree of participation, working with them as partners. These parties share responsibilities with the lead project team in developing the model and in conducting the analysis and also benefit from mutual learning.

Shared Vision Planning Circles of Influence structure collaboration through concentric circles allowing stakeholders with differing levels of involvement to contribute to the collaborative modeling and build trust and understanding. Trust is developed in concentric circles (Figure 24); the core planning team (Circle A) works to earn the trust of the leaders other stakeholders already trusts (Circle B). These representatives of stakeholder groups contribute information to the core planning team and disseminate information about the study to their member groups (Circle C). Disputes that arise during planning are addressed using a range of alternative dispute-resolution methods. The core planning team presents and elicits comment in forums already being used by stakeholders, such as city water advisory boards or boating groups. Ultimate decision makers (Circle D) are engaged in workshop settings to ensure that the collaborative model building addresses the objectives and performance measures that most impact their decision making.

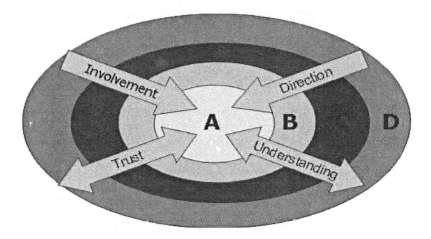

Figure 24. Shared Vision Planning Circles of Influence

The third SVP element, the "shared vision planning model," is the decision support tool that facilitates the identification and evaluation of alternatives that directly address the planning objectives. The primary purpose is to provide a computational environment in which alternatives can be generated, refined, displayed, and evaluated. These models incorporate the hydrology, hydraulics, and physical features of the water resource system and include essential economic, environmental, and social impacts that guide water resource management. But importantly, within SVP, the model is the outcome of stakeholder involvement in their design, development, and use.

Palmer et al. [2007] cite three distinguishing characteristics of SVP models:

- *Customized* - The model addresses specific stakeholder needs and concerns identified in the planning process. This characteristic typically leads to the development of a *customized model* rather than the application of a generic model to ensure comprehensiveness and flexibility.
- *Collaboratively Built* - Planning participants, stakeholders, and water resources professionals jointly construct the model, increasing their engagement and their confidence in the model's results.
- *Highly Interactive* - The model encourages participants to use the model throughout the planning process and ensures that it addresses issues at an appropriate level of detail and accuracy. The act of vigorously exercising the

model allows stakeholders to report concerns about the planning process as well as the model itself.

Conclusions

Thus far, most SVP processes have built models using a generic object-oriented programming package (e.g. STELLA, Extend, PowerSim); however, is it possible to use other types of software (e.g. Spreadsheets, GIS, WEAP). The important point is that the technical analysis is transparent, integrated, and credible (as determined by experts and stakeholders); the needs of the process should drive the choice of modeling tool (Palmer et al. 2007).

In summary, the SVP process combines the best features of more traditional technically based planning processes and consensus-based decision-making processes. SVP is a process that has been successfully applied for two decades and continues to evolve. Since the SVP was codified in the drought study, many studies have employed SVP concepts.

Although there has been notable progress in SVP over recent decades, great opportunities still exist to refine the methodology and expand the use of these ideas. Practitioners who use SVP and other related approaches have identified critical elements and are beginning to refine the SVP methodology in a number of ways. By synthesizing the key principles that define SVP, we can clarify the methodology and support communication among current and new practitioners. Present work will identify performance measures for assessing the benefits of applying SVP techniques and principles. Through comparing SVP concepts with related and overlapping approaches and tools (e.g., conflict resolution, mediation, visualization models), we can reinforce and improve SVP methods. Improvements in both the software and in the public engagement process are possible.

As applications of SVP are now growing rapidly, there is a great opportunity to expand the use of SVP. The products from the methodological refinements will help teach planners and recruit new practitioners. Revised curriculums and training will foster a new generation of planners that are comfortable with applying SVP principles. Specific areas of expansion that are being explored include the use of SVP process within the regulatory setting of Section 404 of the Clean Water Act. Collaborative processes employed during the initial development of water supply plans may avoid later delays from litigation and result in improved water supply alternatives that meet environmental standards (Stephenson 2000).

For the foreseeable future, stakeholder collaboration will play an ever-increasing role in water resources management and computer technology will continue to provide vital decision support tools. Techniques such as SVP can play an important role in merging these two trends. Further research to develop the conceptual foundations of these techniques will help provide appropriate tools for water managers to solve tomorrow's most vexing water resources problems. It is an exciting time in water planning.

References

Arnstein, S.R. 1969. "A ladder of citizen participation". J. of the American Institute of Planners 35(4). 216-224.

Delli Priscoli, J. (1995). "Twelve Challenges for Public Participation". Interact – the Journal of Public Participation. 1(1), 77-93.

Institute for Water Resources. (1994). *Managing Water for Drought*. IWR Report 94-NDS-8. Alexandria, VA.

Lund, J. (2007) personal communication.

Maass, A., M. M. Hufschmidt, R. Dorfman, H.A. Thomas Jr., S. A. Marglin, G. M. Fair, (1962). *Design of Water-Resource Systems; New Techniques for Relating Economic Objectives, Engineering Analysis, and Government Planning*. by Harvard University Press, Cambridge, MA.

Major , D.C., and H.E. Schwarz, (1990) *Large-Scale Regional Water Resources Planning: The North Atlantic Regional Study* (Dordrecht, The Netherlands: Kluwer Academic Publishers, Water Science and Technology Library, Vol. 7).

Palmer, R.A., Cardwell, H.E., Lorie, M.A., and Werick, W. (2007). "Disciplined Planning, Structured Participation, and Collaborative Modeling – Applying Shared Vision Planning to Water Resources Decision Making", ASCE J Water Resources Management and Planning, *in review*.

Palmer, R., (1998). "A history of shared vision modeling in the ACT-ACF comprehensive study: A modeler's perspective," *Proceedings of Special Session of ASCE's 25th Annual Conference on Water Resources Planning and Management and the 1998 Annual Conference on Environmental Engineering*, W. Whipple, Jr., ed., Chicago, IL, 221-226.

Punnett, R. and Stiles, J. (1993). "Bringing people, policies and computers to the water (bargaining) table." *Proceedings of the 20th Annual National Conference, Water Resources Planning and Management Division of ASCE*, Seattle, WA.

Stephenson, Kurt. (2000). "Taking Nature into Account: Observations about the Changing Role of Analysis and Negotiation in Hydropower Relicensing." William and Mary Environmental Law and Policy Review 24:3 (Fall 2000): 473-498.

U.S. Water Resources Council. (1973). *Water and Related Land Resources: Establishment of Principles and Standards for Planning*. Washington, D.C.

South Florida Water Management District Operations Management

Introduction

This final case study takes us back to South Florida, where our case studies began. From the two previous case studies that reference Florida, it is easy to deduce that water resource operations management is a critical issue in the South Florida Water Management District (SFWMD). A steadily growing set of objectives resulting in many diverse requirements for water control has recently been broadened by an intense schedule of infrastructure construction to return the flow regime to more natural conditions. Under these multiple objectives, a customized system of control has evolved.

As discussed in Chapter 4, the Comprehensive Everglades Restoration Plan (CERP) has been undertaken by SFWMD, which is aimed at restoring, protecting, and preserving the water resources of the central and southern Florida ecosystem Through nearly 60 separate civil engineering projects, CERP will restore flows to the ecosystem that are similar to the conditions before the introduction of water control infrastructure (CERP 2000). The Acceler8 program, which has put several of these projects on the fast-track, has intensified the need for new technology to manage water control infrastructure (SFWMD 2006).

SFWMD is governed by a board composed of district citizens. Appointed by the governor and confirmed by the state senate, these individuals collectively represent the full geography of the region. To help the board make decision and carry out policy, the SFWMD employs a large staff of nearly 1800 led by an executive director and an inspector general. The operation of the District's water control infrastructure is conducted by the Operations Control Department, which is one of several departments in the Operations and Maintenance Resource Area.

Current Water Control Infrastructure

The water control infrastructure at SFWMD began as six canals. From this, it has evolved into the current infrastructure, which comprises 1800 miles of canals and levees and 160 major drainage basins. This infrastructure is controlled through 200 strategically placed gates and 27 pump stations. A state-of-the-art Supervisory Control and Data Acquisition (SCADA) system controls 70 of the gates and six pump stations remotely. The remaining structures are controlled manually by local employees (Mierau 2006).

The SCADA system comprises a hybrid wireless and wired network of control devices that are operated centrally in the Emergency Operations Center by the system operators and can actuate the pumps and locks in the system. The SCADA system was installed to ensure control of the infrastructure during critical events such as hurricanes, even when District personnel cannot reach the control structures in person.

The control structures are operated by members of the Operations Control Department at the Emergency Operations Center (EOC), which is located at SFWMD headquarters in West Palm Beach. The EOC is fortified to withstand a Category 5 hurricane and is stocked with supplies and living quarters to support operations staff for an extended period in the event of a severe storm.

An Evolving Set of Solution Requirements

The current water control solution in place at SFWMD is made up of the physical infrastructure, monitoring and control infrastructure, and a custom-developed set of decision support tools (Ryan 2006). This solution can best be described as the result of a process of evolution. Constrained by the need to continually manage the water resources of the region, the solution has grown over the years through alternating injections of powerful water resources theory and sessions of ground-truthing implementation. This process of gradual addition and augmentation has provided many practical lessons and the salient conceptual requirements of the solution have revealed themselves. The major requirements, as communicated by the current operations staff, are listed below. While not a comprehensive requirements analysis, this list communicates the key needs; meeting them has proven most valuable from the perspective of the operational staff (Mierau 2006).

- **Intentionality** – Water-management decisions related to emergency flood control have had serious ramifications and have in some cases lead to legal challenge. The ultimate responsibility of meeting the objectives of the District falls on the shoulders of the governing board. As such, the planning and management of water control infrastructure must be clearly documented and justified, even in cases of emergency where time for documentation is limited. The operational control solution must therefore institutionalize 1) policies of control that have been agreed to and sanctioned by the governing board and 2) in the event of emergency when pre-determined policies will not suffice, recording procedures to track the decisions of the operator

- **Real-time Monitoring and Control** – The requirement for real-time monitoring and control is most clearly revealed in the case of emergency response, where an instantaneous picture of the state of the system is critical for decision making.

- **Centralized Control** – As with real-time control, emergency response underscores the need for a centralized location for control of the water infrastructure. With the risk of storm-related damage to control facilities, a centralized, fortified structure is the best approach for reducing risk. Moreover, decisions made by operators and strategists require conference and consultation. With a centralized control facility, this consultation is possible and most efficient.

- **Scalability** – With the new CERP projects, the system is expected to grow quickly and a much larger set of decisions will have to be made by the operations staff. The system must therefore grow to accommodate the new infrastructure while maintaining the same operational style and efficiency.

Solution Concepts and Design

The conceptual requirements listed above are met through application of state-space, control volume, and rules-based decision support theory. Perhaps the most difficult requirement to meet is that of intentionality, as intentionality is not a single control parameter to be set but is rather a goal that manifests itself in multiple control parameter settings. The following sections will detail the application of control volume, state-space, and rules-based decision support within the context of SFWMD operation. The total solution design will then be explained.

State-space Representation

The theoretical foundation on which the solution lies is knowledge of system state. *State* can describe the health of the measuring devices, the values of measured variables at gauge locations, or more abstract concepts such as water availability for a certain geographic region (Mierau 2003). Operationally, knowledge of the current state of the system is critical for decision making. A state-space approach to describing the system is therefore essential.

Dimensionality

As Figure 25 shows, a single control structure comprises several gates, each of which has an upstream and downstream stage. The electric motors used to move the gates each have operational health indicators. There may be several gauges to determine flow at the structure, including some that are redundant. The number of measurements to be taken at a single structure can therefore number in the tens at least. Since there are more than 200 structures and these represent just a fraction of the measurements of state that need to be recorded in real-time to fully quantify the system state, it becomes clear that the complete control system will be represented by thousands of variables. In state space theory, this system is said to be of high dimensionality because, at a given time, the value of each of these variables can be one of any values in a range. Therefore, to understand the current state of the system and how the system will evolve to the next time step, thousands of variables must be retained. The high dimensionality presents challenges in terms of computational tractability and in terms of ease of understanding on the part of the operators.

The lesson of dimensionality has been learned the hard way at SFWMD. Initial attempts to implement systems that fully characterized all dimensions of system state resulted in slow running times and computer crashing. A solution implemented in the early 1990s was reported to require three hours to simply start up (Mierau 2006).

Parameterization

The curse of dimensionality means that summarizing the state variables to parameters that measure system state in a more abstract way is a requirement (Bellman 1961). Parameterization was introduced to reduce the dimensionality of the system. The parameters are broken into three general categories: control system health, monitored variables, and abstract water resource concepts. The control system health parameters summarize the health of the measuring and control devices within the SCADA system. The monitored variable parameters summarize the measurement of meteorological and hydrologic variables at single monitoring locations. The water resource concept parameters produce a regional- or watershed-based estimate of state. An example would be the average stage (elevation of water surface) within a certain geographic region (Mierau 2003).

Figure 25. A typical water control gate (USACE 2003).

A Control Volume Approach

The third class of parameters - the abstract water resources conceptual class - is primarily implemented via the control-volume based water control system (WCS) concept. Figure 26 illustrates a water control system. The introduction of the WCS concept provided a valuable tool for quantifying the state of the water resources in an intuitive and analytically practical way (Ryan 2006). Similar to a watershed, a water control system comprises land and water bodies, the inflows and outflows of which are managed by control structures. The land portion of the water control system is referred to as the *water control catchment* (WCC) and the water body portion is referred to as the *water control unit* (WCU). The WCS concept has been applied to the entire SFWMD domain and resulted in 168 separate WCSs.

Rules-based Decision Support

With a necessary but tractable set of state parameters, decision support is aided by a rules-based decision-support system. This system sets a time-varying envelope for state parameters and provides operators with warnings when the parameter value moves outside of the envelope. Additionally, these rules can be used to enact predefined operational strategies that will, to the extent possible, bring the parameters back to within-envelope levels and thereby automatically maintain normal operating conditions in the system.

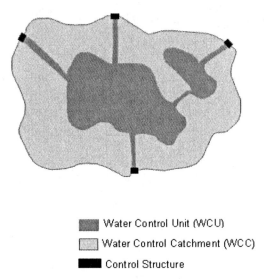

Water Control Unit (WCU)

Water Control Catchment (WCC)

Control Structure

Figure 26. A water control system (Bourne 2006).

The rules-based system is the ultimate tool for meeting the requirements of automatically documented intentionality on the part of the operators. If a rule base is agreed upon and sanctioned by the governing board, the actions of the operators are not only informed by the pains-taking process of developing the rule base, they are documented. Of course, straying from the rule base can occur during extreme periods if improbable conditions occur. But, minimizing the frequency of these occurrences through a rule base that anticipates most events minimizes risk to life and legal risk.

It is important to note that the concepts presented above are dependent and inter-related. To implement the rule-based decision-support design, the system state must be known. For the system state to be known, the calculation of system state must be tractable and thus the dimensionality must be reduced. For the operators to intuitively understand the system, it must be described in an accessible way – the control volume. The solution must therefore integrate these concepts into a cohesive and parsimonious whole to meet the conceptual requirements. The next section describes the solution in place and how these concepts are included in it.

A Three-tiered Solution

The current solution integrates the state-space, control volume, and rules-based concepts into a three-tiered system, as shown in Figure 27. The lower tier comprises the physical infrastructure, which includes canals, levees, control structures, and pumps. The second tier comprises the electronics infrastructure, which is used to monitor and implement control strategy on the physical tier. This tier primarily contains the SCADA system, but it also contains the computer hardware and software necessary for data storage – primarily in relational database management systems (RDBMS). The third tier is the decision-support tier, which analyzes the monitored data streaming from the second tier and provides information to the operators about the state of the system to support decision making. The decision-support tier is also capable of suggesting courses of action and, in the future, will be able to implement them automatically. Decided-upon control strategies are implemented through control infrastructure in the second tier and finally result in a modified physical infrastructure state in the first tier. Because the result of implemented policies is automatically monitored and conveyed back up to the decision tier, a feedback loop is created.

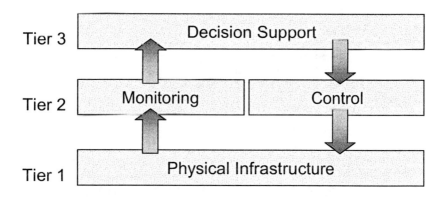

Figure 27. Current Three-tiered Solution Design.

The third tier represents the current operational decision-support system (ODSS). Developed primarily on-site by SFWMD Operations Control Department staff, this system integrates the theoretical concepts learned over the years with the hands-on knowledge collected at SFWMD in the day-to-day operations environment. Still under a prototype classification, this system has been in use operationally for approximately seven years. During that time the system has grown and matured. Development of the rules base has been of primary importance.

The state of the Tier 1 physical infrastructure and natural features is measured by the Tier 2 electronics infrastructure. The state data is analyzed by the decision-support tier and results are communicated to the operations staff. When decisions are made, resulting control policies are implemented in the physical infrastructure through the control electronics infrastructure.

The ODSS is comprised of a graphical user interface (GUI) that makes use of a data model designed to emulate the physical conceptualization of parameters and water control units (Ryan 2006). The GUI captures data from the Tier 2 SCADA and RDBMS via UNIX sockets. The intent of the user interface is to provide a graphical means by which operators can assess the state of the system and compare the state to the target levels, implement control strategies, document operational intent, and configure the system. The GUI was developed for the UNIX operating system with LISP and C programming languages. The main tools contained within the GUI are described below.

Display State Screen

The display state tool was the first part of the solution to be developed. Its purpose is to provide a means of assessing predefined and ad hoc queries into the current and near-past system state. Through the results of these queries operators can create an accurate picture of system state and trajectory for simple to complex parameters.

Review Status Screen

The review status screen provides a picture of the current state of the system and supports graphically driven control of the SCADA system. This tool was developed more recently than the display state tool. It acts as the primary means of control for the operators.

Objective Graph Trigger Manager

One of the most useful operational tools, the objective graph trigger manager, facilitates graphical creation of trigger envelopes for state parameters, which are defined by a high and low monthly series of thresholds defined over the year. If the subject parameter's level is within the envelope, then no warnings are fired. As the parameter's level starts to move outside the envelope, warnings can be fired to prompt operator or automatic control action to bring the parameter back within the envelope.

Typically, several envelopes are tied to various rule sets. These rule sets can be tied to different objectives such as management of water storage area in detention basins or water bodies or reducing severity of departure for measured parameters. If the

110

parameter's level moves outside a specific envelope, that envelope's warnings and consequent actions will be enacted.

Rule Manager

The rule manager is the primary tool used to define the rules for the various parameters. Through an intuitive scripting language, new rules can be devised and implemented. In general, the information required to make a rule includes the specification of the object to which the rule relates, the rule type, and a short description of the rule. The description of the rule specifies the trigger and action sets. The trigger set consists of all zones within an objective graph that must be violated simultaneously to enact the action set. The action set can post/retract objective graphs, generate warnings or alerts, activate/deactivate other action sets or plans, or even issue commands to the SCADA system itself. A plan is a named action set. Plans can be activated manually or automatically by rules or other plans. Figure 28 illustrates the relationship between objective graphs, rules, and actions/plans.

With the toolbox of rules, triggers, and plans that the current ODSS provides, complex operational schemes can be implemented. Operators can construct cascading and conditional system of actions tuned to respond to very specific and complex rules, allowing the operators to use their wealth of experience and instinct. As these complex plans are recorded through the ODSS tools, the understanding of the water control system held only by the operators can be objectively recorded and used for system operation in the future.

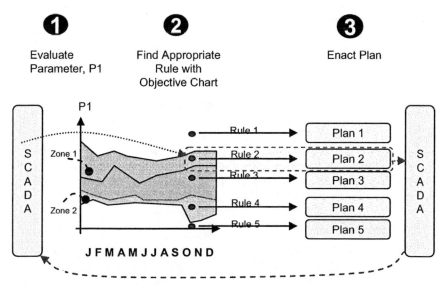

Figure 28. Illustration of rules-based decision-support process. *(Here state parameter, P1, is calculated for October and compared to the family of objective*

graphs that have been developed for P1. Depending on where P1 lands, a rule is triggered (here Rule 2) and an action plan is suggested (Plan 2). Currently, plans send warnings or suggested operations to the operations staff to prompt operator action. In the future, full automation of the SCADA system, where the plan is implemented automatically, will be possible.)

The Future

The future of operations control at SFWMD will build on the existing knowledge base with the primary intent of integrating with other water resources planning and management teams. Chief among future plans is the new operational decision support system (ODSS). In a collaborative environment, the ODSS will be developed with input from several departments at SFWMD, each with interest in ensuring accurate and comprehensive understanding of the District's water resources (Stewart 2006).

In previous implementations of water resources projects, SFWMD has pioneered a collaborative method of agency-wide solution development. With members of SFWMD's Information Resources Department acting as facilitators, business processes are defined with the help of subject matter experts from various disciplines. In this way, the solution that is developed benefits from up-front input from the user base, input that drives the development process (Hampson 2005). This same process will be implemented for the development of the new ODSS.

Departments concerned with the SFWMD's enterprise GIS, regional simulation modeling, biological and ecological modeling, and flood mapping are the main set of groups to be involved. This inclusive approach will ensure the most efficient data flow from the real-time operational setting to the databases used for retrospective analyses that these scientific discipline groups conduct. Moreover, the inclusive approach ensures awareness and accessibility of SFWMD's considerable data stores for the broad user base.

A recently developed tool in achieving the ODSS vision is the prototype GIS-based WCS tracker tool. This tool calculates a real-time water budget for each WCS in the district on a 15-minute basis. The water budget computes the storage of water in the land-based portion of the WCS, the water control catchment. This storage is a valuable descriptor of WCS state as it quantifies on-coming demand on the water control infrastructure. The WCC storage represents a complex parameter that will be fed into the ODSS and drive decision making (Bourne 2006).

Conclusions

SFWMD has the formidable responsibility of ensuring a safe, environmentally healthy, and sustainable water resource that meets the needs of a growing population. Through an evolutionary process, a water control system has been created that both implements powerful water resources planning and management theory and respects the pragmatically built knowledge base of SFWMD operators. The result of this process of evolution is a system that efficiently uses technology to achieve the multiple water resources management objectives of SFWMD.

References

Bellman, R.E. (1961). *Adaptive Control Processes*. Princeton University Press, Princeton, NJ.

Bourne, S., Stewart, K., Mierau, R., Hampson, J., and Brumbelow, K. *"A White Paper on ODSS Tools: WCS Tracker, Conceptual Presentation."* South Florida Water Management District, West Palm Beach Fl.

CERP Staff (2002). *Progress Report Presentation*. Presented at CERP Governing Board Workshop, South Florida Water Management District, West Palm Beach Fl.

Hampson, J.C. (2005). *"Enhanced Arc Hydro for South Florida Water Management District; Task 1.2 Requirements, Definition and Conceptual Design."* South Florida Water Management District, West Palm Beach Fl.

Mierau, R. (2003). "Real Time Water Control Operations" in *"Proceedings of ASCE Symposium, Evolution of Hydrologic Methods through Computers,"* Philadelphia, PA.

Mierau, R. (2006). *Personal Communication*. South Florida Water Management District, West Palm Beach Fl.

Ryan, J.P. (2006). *Personal Communication*. South Florida Water Management District, West Palm Beach Fl.

SFWMD (2000). *2000 District Water Management Plan*. South Florida Water Management District, West Palm Beach, FL.

SFWMD (2006). *Acceler8 Program Website, https://my.sfwmd.gov/portal/page?_pageid=1855,2832596&_dad=portal&_schema=PORTAL&navpage=overview*. South Florida Water Management District, West Palm Beach, FL.

Stewart, K. *Personal Communication*. South Florida Water Management District, West Palm Beach Fl.

United States Army Corps of Engineers (USACE), 2003. Manatee Protection Study Report. http://www.saj.usace.army.mil/pd/manatee1.htm, USACE, Jacksonville, FL.

CHAPTER 6

CONCLUSIONS

Technology has always played a significant role in water resources planning and management and is destined to continue to do so in the 21st century (ASCE 1998; Loucks and van Beek 2005; Viessman and Feather 2006). Over time, however, technology's role has changed as a result of (1) the development of high-speed computational capabilities that can support multi-dimensional modeling efforts and (2) the emergence of direct stakeholder participation in planning processes focusing on non-structural approaches. In a sense, the role of technology has changed from being a driver of water management plans (let's build a dam) to a platform for exploring a broad spectrum of ways to deal with international water problems. Regardless of its role, technology is a powerful and effective tool.

Observations on Case Studies

Many international, national, state, and local agencies have developed customized water resources technologies which play critical roles in the protection of human health and the environment. The Washington, D. C. Metropolitan Area's water supply problem was solved through an uncommon application of modeling techniques combined with the strong support of local government leaders who supported abandoning traditional planning concepts and applying innovative new system-oriented water-management approaches. The solution was based on cooperative systems analysis and management rather than on structural development. Risk-management techniques and simulation and optimization models were employed.

In Texas, a Water Availability Model (WAM) has proven to be of significant value in water resources planning efforts. The WAM, in concert with institutional change in the state, has transformed the state water planning process into one of the most sophisticated and vibrant in the U.S., one that is used daily in industry, government, and academia for a full slate of planning, regulatory, management, and research applications. In Libya, the Great Man-Made River Authority has used optimization models to examine and estimate the costs of numerous combinations of the proposed Great Man-Made River Project and various saline water conversion systems.

As we have seen, Florida has several good examples of significant water resources technologies. The South Florida Water Management District (SFWMD) has developed state-of-the-art modeling tools to support the Everglades Restoration Plan and the SFWMD's other programs related to water supply, natural systems, and floodplain management. The SFWMD's efforts to create a comprehensive Operational Decision Support System provide a wealth of lessons. Efforts to restore the Kissimmee River system have included the application of physical modeling techniques to evaluate alternative restoration scenarios. Modeling approaches were important and contributed to the final selection of a restoration plan. Technology will certainly continue to play a critical role in the protection of human health and the environment of South Florida.

In 2004, the Louisiana Coastal Area Study proposed a $2 billion plan for initiating sustainable ecosystem restoration for the entire Louisiana coastal zone. Technology played a major role in the development of the restoration plan. The plan narrowed knowledge gaps that existed with respect to the functioning of coastal wetland ecosystems and the linkages between hydrology and hydraulics.

In the West, the State of Colorado is developing and implementing a Decision Support System (DSS) to serve several levels of government and the private sector. The DSS spans planning and operational functions and promotes intergovernmental coordination and shared vision planning for the South Platte River Basin.

In Canada, a Shared Vision Model (SVM) was developed to integrate all data generated in the Lake Ontario study. Using this model, various regulatory plans could be run through an evaluation process and the results compared between interests and locations. The model integrated plan formulation and evaluation so that new regulation plans could be designed and immediately evaluated. The SVM enabled all involved to understand how actions taken to affect one part of the system or one interest affected all other stakeholders.

A review of the case studies shows a transition (roughly 1970 to 2007) in the primary role of technology from one supporting structural design to one supporting a variety of analytical tools. These tools provide an integrated analysis of natural systems and in some cases facilitate stakeholder involvement.

This discussion of the current state of technology for water resources planning and management is neither complete nor meant to imply that what we have does the job to the degree we would like. It does not. There is plenty to do to improve our planning and management technology, even if our computer and communication hardware capacities remain constant—and surely they will not. It is just as hard for us to imagine what will be developed and available for planners and managers in the next half century as it was for the Harvard Program participants some 50 years ago to imagine what technology we would have today. But we can take a stab at what seems like reasonable opportunities for further improvements in our technology.

We need better ways of accomplishing the following:

- Readily identifying the location databases containing needed data and being able to download those data and then convert them, without user involvement, to the format required by any model. This includes data from geodatabases.
- Creating more realistic displays of model outputs, such as overlaying model results onto video displays of the system and to be able to fly over (or even under) the video displays of the system to view the system at any location desired.

- Using different types of optimization as well as simulation algorithms, as desired, without user concerns about formats and approximations needed by the solution algorithms.
- Performing real-time simulations and optimizations within a Virtual Reality (hologram) environment at reasonable costs, or at least an approximation of that on a computer display. This might be especially useful in simulations of natural disasters or for training in educational institutions.
- Obtaining on-line help via audio/video links as well as email when help using some analysis method or software package is needed.
- Developing models useful for aiding negotiation processes as well as for understanding the system and issues being negotiated.
- Using the resources of the Internet more effectively, including innovations such as Google Earth and models having their interfaces on the Internet.
- Linking our model data bases to remote sensors out in the environment measuring in real time events that are taking place that are of interest in real-time or adaptive management strategies.
- Calibrating and verifying model parameters and performing sensitivity and uncertainty analyses.
- Developing software that makes creating DSSs of specific water resource planning and management problems more efficient, more effective, and easier to use in shared-vision modeling exercises.
- Educating, training, and inspiring the next generation of modelers and empowering these future managers and analysts to use technology to the greatest extent possible.

General Comments

Technology enhances the ability of water resources planners and managers to achieve their objectives, but it should not be considered the driver. Rather, technology should be recognized as a strong and versatile tool for evaluating alternatives and their environmental, social, and economic impacts. Unfortunately, predicted impacts are never certain, and the natural and social sciences offer no help in determining the best course of action to take when confronted with the conflicting goals of multiple stakeholders.

The notion that so-called "best technical solutions" are the ones to be followed is no longer valid. What should be done may not even be clear, but understanding the social-political-economic setting can suggest options that will not be acceptable and should not be further considered. In most cases, technologists are only one of a diverse group of stakeholders having an interest in and contributing to the management of water. Planning is not just the application of science. It requires creating a social environment that involves, at the outset, all of the key players in a dynamic process. Engineers and planners must learn how to plan and design in the context of the reality of the world at the moment. The primary conclusion is that technology is a valuable decision support tool, but it is not the decision maker.

Whether proposed water resources policies and plans are implemented depends on their acceptability to decision makers. Final decisions are made politically, but they can be influenced by analyses made by water resources planners and researchers. The extent to which such studies are considered depends on the credibility of the analyst, an understanding of the political and social climate of the planning region, and the directness of stakeholder input.

The analytical capability at our command is significant, and in the future it will likely expand beyond our current imaginations. Unfortunately, our ability to use these analytical tools to address the "real" dimensions of the problems we face is constrained by political, financial, social, agency, legal, topical, and other boundaries. Many of these boundaries have long been institutionalized and are difficult to change. To deal with this problem, we need forums that can address true dimensionality and that can be divorced from those that cannot. For example, government agencies, consulting firms, non-governmental organizations, and citizens typically have particular limitations, authorities, and expertise. Their agendas are often in conflict with each others'. Such circumstances directly constrain achieving an integrated approach to water resources planning and management and indirectly constrain the application of technology in the true spatial and temporal context of the problem to be addressed.

We need to get out of the traditional mentality of dealing with problems within the dimensions of groundwater, surface water, water quantity, water quality, city boundaries, county boundaries, state boundaries, national boundaries, government and agency authorities, regulatory policies, and traditions, for example. Problems should be solved in the context of their true spatial and temporal boundaries. But extricating from the "box" constraint is easier said than done. It can be achieved, however, if the right social and political environment is created.

The Washington D. C. Metropolitan Area Water Supply case study is illustrative. Historically, the three principal water supply agencies had been operating their systems independently, concerned only with what they considered best for their constituents. Taking a regional perspective was not considered.

Then, in the late 1970s, the political and institutional leadership in the WMA, facing urgency in solving their water supply problem, agreed to support the exploration of imaginative and unbounded non-traditional technical options. The USACE, the states of Maryland and Virginia, the District of Columbia, the Interstate Commission on the Potomac River Basin, the Fairfax County Water Authority, the Washington Suburban Sanitary Commission, the Metropolitan Washington Area Council of Governments, and other stakeholders provided the forum for coordinating their water management policies. Research conducted to support the objectives of the forum disclosed that if the three WMA utilities were operated as a system, and if releases from the Jennings Randolph Reservoir were made daily, only a small portion of the storage originally proposed by the USACE would be needed (Hagen et al. 2005). It was also found that by operating the WMA water supply system as a whole (using all facilities regardless of ownership) the water supply requirements of the WMA could be met until about 2025 (Sheer 1981).

This unconventional union of institutional cooperation and technical expertise resulted in a solution to the water supply problem of the WMA that had been sought unsuccessfully since the 1950s. A succession of studies had terminated with a recommendation that the only way to meet the future water supply needs of the WMA would be to construct two large reservoirs. This belief was proven false when analysts discovered that operation of all existing reservoirs and utilities as a single coordinated system would eliminate the need for additional storage for long into the future (Hagen et al. 2005, McGarry 1990, Sheer 1981).

It is worth noting that the U. S. Water Resources Council (WRC), established by the Water Resources Planning Act of 1965, provided a mechanism for unbounded consideration of water issues at the national and regional level. This capability was eliminated in 1982 when the Reagan Administration zeroed out the Council's budget. As of this writing, proponents of restoring some form of analyzing and coordinating body such as the WRC have been unsuccessful. But the need for such an institution at the highest level of government is recognized.

Our extraordinary analytical capabilities offer enormous decision-support capability, but if they are not applied to the true dimensions of the problem being addressed, the solutions achieved will not be the best that could have been generated. Efforts to relax constraining influences must be assigned a high priority.

Expectations for the Future

We believe that the role of technology in water resources planning and management will broaden in the future, and this role will be characterized by the following features:

- Integrated water resources planning and management will become more widely accepted as the goal. This will require applying technology in the context of the true "problemshed" and addressing the unbounded linkages among physical, spatial, temporal, environmental, social, and institutional dimensions.
- Emerging new modeling and monitoring technologies combined with faster generations of computers will continue to expand our analytical capability.
- Feedback from monitoring programs combined with an ever-increasing ability to measure interactions of ecosystem components will lead to a better understanding of the quantities of water needed to support these systems. This will, in turn, provide an improved baseline for allocating limited water resources among all competing uses.
- Advances in nanotechnology will foster new options for dealing with contaminated waters. This could significantly affect our ability to carry out treatment of groundwater in situ.

- Universities will partner more effectively with state, federal, and local government agencies; non-governmental organizations; consulting firms; and others in developing research needed to support water resources planning and management. Such partnerships will enhance academic understanding of practitioners' needs for tools to more effectively carry out their missions and also foster translation of new theories into practices that can be incorporated in planning and management processes.
- Visual displays of model outputs will be designed to more easily accommodate the needs of decision makers. Such video displays will enable the viewer to fly over or under a system of interest and to view it at any desired location.
- Analytical models will be more widely used to support negotiation processes and to enhance understanding of the system and issues being negotiated.
- Increased partnering and stakeholder involvement in water resources planning and management will foster expansion of shared vision planning and modeling approaches.
- Climate change models will become more definitive and will become widely integrated into water resources planning processes.
- Objective forums for addressing the true dimensions of water management problems will become more common and they will support broader application of analytical techniques.
- The concept of sustainability will become widely incorporated into water resources plans, and models to compare levels of sustainability among alternatives will be needed.
- Real-time and adaptive management strategies will be supported by model data bases that are linked to remote sensors measuring events in real time.
- Interdisciplinary and non-linear approaches to water resources planning and management will become the norm, shifting away from traditional linear and engineered approaches.
- The use of the internet, geographic information systems, global positioning systems, and personal computers will expand and broaden in scope.
- New water resources planning models will be developed and applied as effective problem solving tools. Existing models will be refined and more widely used.
- Real-time simulations and optimizations will be performed within a Virtual Reality (hologram) environment on a computer display. This will support the study of natural disasters and serve as an educational tool.
- There will be an increase in regional and river basin planning, and it will heavily depend on analytical capability.
- The need for interstate and international water management models will accelerate, and technology will play a major support role in their development.

Final Thoughts

Increasingly, those involved in water resources planning and management recognize the importance of establishing a more open and participatory decision-making process. This need also points to the need for coordination among the many water institutions that govern water resources and stakeholders that are affected by management decisions. This coordination then motivates improved analytical modeling tools, such as decision support systems, that can support consensus building and resolve disputes.

Today, experts recognize that water resources are a part of numerous complex natural and social systems. Advocates of integrated resources planning (encompassing, for example, water, energy, and land-use planning) make a similar point. These perspectives present numerous intellectual, analytical, and evaluative challenges. In making policy choices, trade-offs among competing values are inevitable.

Getting stakeholders involved in developing their own models has been shown to be an effective way of building a consensus—a shared vision. Accomplishing this will take more than just good modeling building shells into which participants of a model building exercise can draw their system and enter its data. It will also take some instruction from those of us who create the tools that can be used for these exercises. We need serious thought about how such modeling tools should be developed and how we can best get all interested stakeholders involved in a particular decision-making process to use these tools, especially when stakeholders view the world quite differently. If we can actually get all of the decision makers to use these tools, it might have more of an impact on water resources decision making than all of our models have had to date. A futuristic scenario written by Dr. Daniel P. Loucks is presented in the Appendix that follows. We hope that our readers will enjoy this imaginative exposition.

References

ASCE.Task Committee on Sustainability Criteria for Water Resource Systems (1998). *"Sustainability criteria for water resource systems,"* American Society of Civil Engineers, Reston, VA, ISBN 0-7844-0331-7

Hagen, E. R., Holmes, K. J., Kiang, J. E., and Steiner, R. S. (December, 2005). "Benefits of iterative water supply forecasting in the Washington, D.C., metropolitan area," *J. of the American Water Resources Association (JAWRA)*, Middleburg, VA, 1417-1430.

Loucks, D. P. and van Beek, E. (2005). *Water resources systems planning and management: an introduction to methods, models and applications*, UNESCO Publishing, ISBN 92-3-103998-9.

McGarry, R. S. (1990). "Negotiating water supply management agreements for the National Capitol Region," in *Managing water-related conflicts: the engineer's role, edited by Viessman, W. and Smerdon, E. T.,* ASCE, 116-130.

Sheer, D. P. (Nov. 12, 1981). "Assuring water supply for the Washington Metropolitan Area, -- twenty-five years of progress," in *"A 1980s view of water management in the Potomac River basin," Report of the Committee on Governmental Affairs,* U. S. Congress, Senate, 97th Congress, 2d Sess., U.S. Gov't. Print. Off., Washington, D.C.

Viessman, W. and Feather, T. D., Editors. (2006). *"State water resources planning in the United States,"* American Society of Civil Engineers, Reston, VA, ISBN 0-7844-0847-5

APPENDIX

A FUTURISTIC SCENARIO

The sky was gray and a light drizzle accompanied Jos and Nicki when they arrived at the Bath Harbor Regional Environmental Monitoring Center. Eric and Tineke, operators on the night-shift, were happy to see their relief and reported no unusual conditions in the region. They were tired and wanted to get a good meal and some sleep. The Center was staffed 24 hours per day, 7 days a week. The operators were responsible for ensuring that the network of remote and in-place sensors, communication links and computers for converting monitored data to information and knowledge, and for communicating that knowledge to those who needed to know, were properly functioning. Operators had to keep aware of the current and forecasted states of the region's water resources, environment, and ecosystems. They were also responsible for planning the upgrading of the Center's capabilities and facilities.

Should any of multiple environmental indicator and index threshold values be exceeded, automatic alarms would alert the Center operators and the appropriate agencies responsible for taking management actions. Like all shift operators, Jos and Nicki were prepared to provide additional and more comprehensive information obtained from the center's displays in response to any telephone calls or email messages they might get from those management agencies. They also sent real-time data and information to various university research faculty involved in the development of improved predictive models, networks of sensors, analysis algorithms and their computer software.

Jos and Nicki completed the "hand-over" from Eric and Tineke and then queried the master monitoring and display computer, jokingly named Chaos (The goddess of void or emptiness from which all things emerged), to summarize the state of the region. A colorful map appeared on the 4-foot by 6-foot flat display screen on a wall of the monitoring room. In her familiar soft voice, and pointing to the applicable parts of the computer display screen, Chaos began to review conditions of special interest, always listening to see if she should proceed, or alternatively, to answer questions or provide any more detailed information on a particular subject. Hearing nothing, Chaos continued.

When Jos or Nicki asked if the soils in the region's watersheds were saturated due to this drizzle, Chaos estimated it would take another day before that condition would be reached at the present rate of precipitation. If such a condition were reached and were accompanied by increased rainfall, local flooding and landslides might occur. The display showed just where those events might take place. The briefing from Chaos continued, together with numerous displays (using the relatively old fashioned 8 bits or over 16 million different colors) illustrating what she was presenting.

At the end of the briefing on the state of the natural environment in the region, Nicki asked for a status report of all the sensor networks. Chaos asked her if she wanted the good or bad news first. Clearly there was some work to do to fix some of the biorecorders and mechanical streamflow monitors, and a few solar panels needed replacing. Chaos also offered some ideas on how certain software could be improved. Jos told Chaos to send those ideas to Dr. Beek at the local university, since he was instrumental in developing what was now being used. That order was executed before Jos had completed his sentence.

Chaos's biggest, and often repeated complaint, was the lack of sufficient sensors to reduce to acceptable levels the uncertainty associated with some of the estimates of ecological indicators. Chaos even printed out a proposed draft of a proposal to NASA for such funding. Both Jos and Nicki said they would keep trying to obtain such funding and suggested Chaos consider preparing a proposal to NSF to obtain funding via the environmental (CLEANER, CUAHSI, GEON and NEON) observatory initiatives. That suggestion took Chaos an hour, only because of a delay in contacting one of the federal government's computers still operating under DOS (to ensure maximum security against terrorism).

After the beginning-of-shift briefing, Chaos asked Jos if he had stopped smoking yet. Nicki responded by saying Jos' clothes smelled like stale cigars. Chaos agreed, saying that is why she had asked the question. Jos dutifully went to his locker and changed into his lab coat.

One of the benefits of living in Bath Harbor was that it bordered the ocean. Many enjoyed the coastal beaches. This meant that some monitoring of near shore conditions was desirable. Lately there seemed to be an unusual amount of dead fish washing up on the beaches, and Nicki wondered what might be causing it. While nothing unusual was showing up on the monitoring system, she decided to contact Amphitrite, the national computer network located at Silver Springs, Maryland, for monitoring the broader sea and ocean environments. That proved to be a wise decision.

Amphitrite told her that soon people on the beach might notice themselves coughing, sniffing and having itchy teary eyes. (Interestingly the next day just such reactions were reported in the local newspaper.) This would be due to a species of algae that produce a toxin that causes respiratory problems and attacks the central nervous systems of fish, birds and sea mammals. Amphitrite went on to explain that blooms of these algae are called red tides. Even when not visible, these algae can kill sea life and taint the air. They can create dead zones devoid of oxygen and marine life. They move around with the ocean currents and can be hard to notice unless concentrated. If detected early when small, ozone and phosphatic clay can kill it. Amphitrite offered to assist in identifying where such actions could be applied using its automated detection sensors along with satellite photographs. Nicki thanked Amphitrite and sent the information to both the Bath Harbor Environmental Protection Department and to the local Chamber of Commerce, thinking that the potential adverse impact on tourism might be of interest to them and hence help generate the political support to take action before it is too late to do anything useful.

She also told Chaos to be on the look-out for this species of algae. Chaos dutifully repositioned its mobile biosensors to better detect such concentrations near the beach.

Meanwhile Jos was thinking about how the detection of fires in watersheds could be improved. He was a firm believer in thinking about hazards that might occur in dry conditions when everyone was complaining about it being too wet, and vice versa. He wanted to be prepared. Putting out forest fires is expensive, and if a natural or man-made fire can be detected soon enough, the cost of fire suppression can be reduced considerably. This cost savings had to be compared to the cost of more sensors, their security against vandalism, and their associated computer networks.

Based on what Jos read about monitoring buildings subject to arson, he was considering a three-pronged electronic surveillance monitoring system. The three types of monitors would include flame detection devices, infrared cameras, and fiber-optic sensors. Each system would monitor the forest independently and transmit data wirelessly to a central computer system. The computer system could then transmit data to the Bath Harbor Regional Environmental Monitoring Center and the local fire departments. The local agricultural and watershed organizations would also be notified of a potential fire event together with its extent and location.

The flame detector monitors both the infrared and ultraviolet signatures and watches for the characteristic flickering appearance of a flame. Since this device could send false signals due to dust, animal activity or weather conditions, the use of two other components would help confirm if there really is a fire.

The fiber-optic sensor will monitor for changes in temperature, which causes the wavelength of light within the glass fiber to change proportionally. The thermographic infrared camera will track the infrared signature from hundreds of adjacent pixels viewed by the camera's lens, measuring the difference in temperature between individual pixels. A key challenge here involves the information technology side of the system – making sure the pattern recognition software that collects and transmits the infrared camera's data can distinguish between a threatening fire and someone simply lighting a cigarette or better yet, a cigar, thought Jos. How big a fire does it take at a certain distance to really trigger that as a fire event? Jos thought maybe some university egg-heads might be able to help him address this and other interesting questions. But, he wondered, how long would it take, especially if graduate students are involved?

Jos called Bob Batterman, his favorite university professor, not because of his research on ecological and environmental modeling, but because they played squash together each weekend. Bob said he thought the fire detection issue should be coupled with a broader suite of applications, such as the reduction of illegal dumping of hazardous wastes in forests. Video monitors hidden near those sites linked to the same communication network might eliminate such violations. Other sensors in the forest could be added to monitor the conditions that lead to the death of white spruce and sugar maple, to monitor the spread of various invasive species, and to sample the ground and surface water quality to a greater degree. Bob suggested these and other issues related to more comprehensive watershed monitoring and analysis warranted some serious discussion, and volunteered to round up some capable and hungry researchers and come to the Center for an open brain-storming discussion. Jos called together additional Center personnel and set up the conference room for the discussion that afternoon. Jos coerced Nicki into purchasing the necessary cookies – traditional for any seminar involving university personnel – during her lunch break. (He paid the bill.)

Mark, an ecological modeler, began the seminar discussion by talking about how ecological modeling is becoming an increasingly important tool for uniting biological observations with remote sensing and ground-based data networks for improved prediction, resource management and protection of human health. He reminded everyone of the availability of over three petabytes of data from the NASA Earth Observing System Data and Information System (EOSDIS) are helping researchers like him develop earth science applications for decision support. Doing this for the Bath Harbor Region would be possible if he were to be able to obtain data on parameters as diverse as land cover, water surface temperature, dissolved oxygen and nutrient concentrations, precipitation, species distribution, and disease occurrence.

John, the computer scientist in the group, reminded everyone that before environmental and ecological prediction models can be applied to a particular region one needs to convert the large volumes of data from satellite and in-place sensors to meaningful information. Mark then went into some detail explaining how ecological models provide a useful framework for data integration, and they are a key component of the developing capability to generate ecological forecasts. Advances in computing have also increased the ability of ecological models to handle large volumes of data and to explore changes in ecosystem characteristics at a wide range of spatial and temporal scales. But there are still challenges associated with habitat suitability or niche and energy balance modeling, geostatistical pattern and distribution modeling, and data assimilation modeling approaches for predicting the response of populations and biotic communities to different "if-then" environmental change scenarios.

He reminded everyone of The Terrestrial Observation and Prediction system (TOPA) and the Carbon Query and Evaluation Support Tolls (CQUEST) at NASA Ames Research Center as two examples of this type of modeling framework. TOPS automatically integrates Earth observation data from a wide range of sensors to run land surface models in near real time. Using TOPS, scientists at NASA Ames currently produce a comprehensive suite of over 30 variables describing land surface conditions at a variety of spatial scales, while CQUEST provides policy makers and land use managers with predictions of carbon sequestration throughout the US.

Jennifer from the University's School of Public Health felt that everyone was ignoring human health benefits from more intensive sensing. She explained how such data together with landscape epidemiological models can help identify and characterize the environmental role of disease systems and predict the time and location of outbreaks or case clusters. Geographic distributions of vector-borne diseases are dependent on the habitat suitability of the vectors, pathogens, and animals involved in the transmission cycle as well as the proximity to and interaction with human populations. She gave examples of Schistosomiasis in Kenya, West Nile virus risk in New York City, and Lyme disease in Connecticut. She offered to work on the key factors in the successful development and use of such models that require real time high-resolution satellite data, a well designed biological monitoring network, and case history data for disease occurrence.

The meeting ended with a discussion, lead by John and Mark, of the need for better ways of 1) linking data of different resolutions or scales for different disciplines or models, 2) linking agent based models to landscape data and community-level or trophic models, 3) defining the uncertainty in models, and 4) communicating uncertainty in risk estimates derived from model outputs such as disease outbreaks or wildfire risk.

Everyone attending the meeting was impressed as to the range and depth of discussion and vowed to hold regular research discussions in the future, both at the University as well as at the Center. They also got to know others having similar and related interests who they had not known before. The university participants left talking enthusiastically about the establishment of a multidisciplinary center for research in environmental sensing and monitoring and analysis. Long overdue, thought Jos and Nicki.

Following the tradition of MIT, a Center for Research in Environmental Monitoring Systems was created within a week, and an announcement was immediately posted for its first seminar:

An open discussion of the use of
Bioreporters and Biosensors
for
Environmental Monitoring and Management

Background:
Microbial ecology plays a central role in every aspect of life. Their small size and variety of species makes them difficult to study. Bioreporters and biosensors have been developed to study community establishment and interaction. Bioreporters are gene sequences that produce a product that is easy to detect. Biosensors are biological molecules that respond to the presence of an analyte by producing an effect that can be detected by electronic means. Both of these rely on biological sensing of an environmental change, and the production of a detectable change and are non-invasive ways of determining the spatial and time components of the system being investigated.

Bioreporters have been used to detect heavy metals in the environment. Bioluminescent bioreporters are currently being used in site pollution cleanup with bioremediation. Biosensors containing bioreporters can be placed in soils or water bodies to detect advancing plumes of chemicals. Biosensors have been used to monitor pollution, human therapeutics, testing materials to be placed in landfills. They can monitor biological processes in wastewater treatment plants and water bodies.

Sponsored by the University Center for Research in Environmental Monitoring Systems

Refreshments, 4:15 pm,
Discussion: 4:30 – 5:30
Thursday, Room 311, Holling Hall

Come prepared to learn and contribute to the identification of research needs and opportunities.

When Jos and Nicki saw this announcement later that week they felt proud. They and their Center had actually provided the seeds for the formation of a university center, no less! Jos also silently wondered if some sort of biosensor could detect the damage to his lungs caused by smoking an occasional cigar. In any event this collaboration between the two Centers could only benefit both Centers.

Now, back to the day at hand:

Later that afternoon Chaos asked Jos and Nicki if they were paying attention to the soil moisture and streamflow conditions. The rain had been continuous and increased in intensity. When Jos said they hadn't Chaos displayed a map showing the soil moisture and stream flow data, and predicted that unless the precipitation ended the alarms might ring in the bedrooms of the water utility personnel responsible for managing the water resources in the region. Chaos was basing her forecast on observations of river inundation areas, water levels and flow variability from wide-area optical sensors in satellites. These data together with topographic data and hydrologic/hydraulic models were being used to estimate river discharge changes. This was backed up with microwave sensors (such as MODIS) for detecting the onset of flooding in response to intense or prolonged rainfall and water vapor and upper air temperature data from sensors that use tunable diode-laser spectroscopy, attached to over-flying aircraft. Jos recalled how this technology had replaced the need to launch weather balloons every 12 hours as was the custom years ago.

Given this news from Chaos, Jos and Nicki contacted the appropriate authorities that might be interested in the possibility of local flooding. After viewing the displays sent to them from Chaos, a flood warning was announced. [That evening the local TV network showed dynamic animated displays of possible flooding together with their probabilities, all provided by Chaos.]

Soon after completing those telephone calls, an alarm on the Alert Panel for the Sub-region called Gorge flashed red. Noise correlations between four sets of geophones placed in the steep slopes of the south side of the gorge indicated faint activity signals indicative of soil movement. This information was confirmed by the filtered information being provided by the recently installed wireless networks of self-organizing open-source smart sensors, or Motes, together with their TinyOS [operating system] and TinyDB [database]. Moments after processing this information, Chaos, this time in a slightly more stressful voice, assured Jos and Nicki that she had already contacted the local police and that they were in the process of detouring traffic around the normally busy Route 13 that is at the bottom of the south slope in that area. Chaos used this opportunity to complain about the work load she was having to carry that afternoon, and asked when the sun would appear. Nicki suggested she contact Iris who always seemed to predict the local weather better than the meteorologists hired by the local TV station. Chaos responded that she already had, but wasn't happy with the answer. Iris piped up by reminding Chaos that facts are facts, nothing else, and are not altered by Chaos' desires. Nicki told Chaos and Iris to keep their arguments to themselves.

Jos and Nicki's day shift was practically over. It had been a busy day. Jos had intended to become more familiar with the newest in automated and continuous, remote and in situ sensor technology for monitoring, in outdoor environments, the biological, chemical and physical properties of the atmosphere, the landmass and in water bodies, including the detection of pollution and toxic chemicals, and the changing states of various species in ecosystems. He was particularly interested in complex microanalysis systems known as 'lab-on-a-chip' technology involving advanced microengineering, including microfluidics, electrokinetic manipulation and sensor techniques. He decided to take some reading material about those topics home with him to supplement what he could find using Google on the Internet. He did this knowing full well his wife would complain of lack of attention. Jos like Nicki really enjoyed his work, and believed he was helping protect and manage the environment for the betterment of all those who lived in Bath Harbor.

Index

model 56--57; early modeling 52--53; Everglades landscape model 55--56, 55*f*; Everglades screening model 54; hydrologic simulation engine 60; management simulation engine 60; models 52--58, 53*f*, 55*f*, 57*f*, 60--63; natural system model 55; next generation modeling 60--63; operational planning 58--60; regional simulation model 60; restoration 51--64; river of grass evaluation methodology 56; role of models 57--58, 57*f*; South Florida Water Management District 52*f*; South Florida Water Management model 54

Everglades landscape model 55--56, 55*f*

Everglades screening model 54

Fairfax County Water Authority 14; low-flow agreement 16--17; Potomac River water withdrawals 15

FCWA: *see* Fairfax County Water Authority

FEPS: *see* Flood and Erosion Protection System

Flood and Erosion Protection System 92, 93--94

floods: control 43--44; Flood and Erosion Protection System 92, 93--94; floodplain elimination 45; models 8

Florida: across trophic level system simulation model 56; Central and Southern Florida Flood Control Project 43--44; drainage of 42, 44; Emergency Operations Center 105; Everglades landscape model 55--56; Everglades restoration 44, 51--52; Everglades screening model 54; flood control 43--44; hurricanes 43; hydrologic simulation engine 60, 61*f*; Lake Okeechobee and Estuary

Recovery 51; management simulation engine 60; models 53--57, 60, 61*f*; natural system model 55; Okeechobee Flood Control District 43; regional modeling 53--57; regional routing model 54; regional simulation model 60, 61*f*; river of grass evaluation methodology 56; soil subsidence 44; South Florida Water Management District 44, 52*f*, 104; South Florida Water Management model 54; Water Resources Development Act of 1992 44

geographic information systems 6, 9, 28

GIS: *see* geographic information systems

GMRP: *see* Great Man-Made River Project

Great Man-Made River Project 33--41, 33*f*; cost-effectiveness analysis 36--37; demand analyses 38--39, 38*f*; Great Man-Made River Authority 33; modeling approach 37--40; obtaining data 37--38; pipe transport 34*f*; sustainability 40--41; water cost per source 39*f*

groundwater: in Colorado 80, 82; development 32--41; Great Man-Made River Project 33--36, 33*f*; models 86; rights 24; surface-groundwater hydrologic models 84

Harvard Water Program 7

HEC-GeoHMS 28

HSE: *see* hydrologic simulation engine

Hurricane Katrina 73--75

Hurricane Rita 74--75

hurricanes: in Florida 43; Hurricane Katrina 73--75; Hurricane Rita 74--75

hydroinformatics 11

SAMSON 84
Savage reservoir 17
SFWMD: *see* Florida, South Florida Water Management District
SFWMM: *see* Florida, South Florida Water Management model
Shared Vision Planning 93--94, 97--103; applications of 102; characteristics of 101--102; Circles of Influence 100--101, 101*f*; collaboration 100--101; decision support systems 99--100, 101; defined 99; elements of 99; history 97--98; public involvement 100; software 102; transparency 100
shore protection 67
simulation engines: hydrologic 60, 61--62, 61*f*; management 60, 62--63; *see also* models
Sixes Bridge reservoir 14--16
South Florida: *see* Florida
South Platte Decision Support System 85--86; development phases 86; funding 85; groundwater model 86; planning tools 86
South Platte River basin 78--88; modifications 82; North Platte confluence 82, 83--84; precipitation 83; recreation 83; reservoirs 83; snowpack 83; South Platte Decision Support System 85--86; water resources data 85; water supplies 83
SPDSS 85--86
SRM: *see* St. Lawrence River model
St. Lawrence River 88--96, 89*f*; *The Boardroom* 93; data management 94, 94*f*; flow regulation 89--90; International Joint Commission 89; models 92--94; plan formulation 92--94; plan selection 95; policy review 91--92; STELLA 93--94; Study Board 91--92; water level 89--90
St. Lawrence River model 92, 93--94
statistical hydrologic model 93

STELLA 93--94
storm surge 8
SVP: *see* Shared Vision Planning

TCEQ 24
Tennessee Valley Authority 3
Texas 23--32; Brazos River basin 27; drought 24--26; flow naturalization 27--28; groundwater rights 24; municipal water demand 23; riparian rights 23; Senate Bill 1 24--26; surface water availability 26--27; surface water rights 24; Texas Commission on Environmental Quality 24; Texas Parks and Wildlife Department 24; Texas Water Development Board 24; water availability models 26--30; water planning regions 25, 25*f*; Water Rights Analysis Package 28--29; water usage 23
Texas Commission on Environmental Quality 24
Texas Parks and Wildlife Department 24
Texas Water Development Board 24
The Green Book 4
TPWD 24
TVA 3
TWDB 24

U.S. Army Corps of Engineers 3, 7
U.S. Bureau of Reclamation 3, 7
USACE: *see* U.S. Army Corps of Engineers

Verona reservoir 14--16

WAD 14
WAMs: *see* models, water availability
Washington Aqueduct Division 14
Washington Suburban Sanitary Commission 14; low-flow agreement 16--17; Potomac River water